STUDENT SOLUTIONS MANUAL
TO ACCOMPANY

Anslyn & Dougherty's

Modern Physical
Organic Chemistry

Michael B. Sponsler
Syracuse University

Eric V. Anslyn
University of Texas at Austin

Dennis A. Dougherty
California Institute of Technology

University Science Books
Mill Valley, California

University Science Books
www.uscibooks.com

About the Cover -- Taming Cyclobutadiene: An object of physical organic
investigations for decades, cyclobutadiene was finally "tamed" in 1991, when
Cram and coworkers generated the molecule in the cavity of a hemicarcerand.
This supramolecular complex allowed full characterization of cyclobutadiene,
including recording its NMR spectrum at room temperature. See Section 4.3.3.

Text Design: Mark Ong
Cover Design: Bob Ishi
Compositor: Michael Sponsler
Printer & Binder: Victor Graphics

This book is printed on acid-free paper.

ISBN 978-1-891389-36-8
Library of Congress Control Number: 2005903713

Printed in the United States of America
10 9 8 7 6 5 4

Contents

To the Student

This *Solutions Manual* provides solutions (not just answers) to all end-of-chapter exercises in *Modern Physical Organic Chemistry*: nearly 600 solutions, not including multiple parts. Used properly – to compare with your own solutions – this manual will contribute tremendously to your understanding of the concepts and methods presented in the textbook. Used improperly – before you have tried the exercise on your own – its value will be marginal.

Learning physical organic chemistry, like other areas of chemistry, requires much more of you than memorization of facts. You will be expected to learn principles and ways of thinking and to apply them in various contexts to show that you can make sense of an observed product, rate constant, or pK_a value. You will also be expected to use your knowledge to make predictions and design experiments to test your predictions. Such skills cannot be learned by reading someone else's answer. You might recognize that it makes sense, and you might pick up another fact or two, but you will not have gained the valuable experience of working through the issues on your own!

Like the textbook, this *Solutions Manual* has "GOING DEEPER" highlights on selected issues – 22 in total. These are provided to explore intriguing issues that go beyond the question that is posed in the exercise. We encourage you to develop the habit of "going deeper" when you come across an interesting question that is not so simply answered or a question that leads to more questions. Physical organic chemistry is full of such opportunities to be inquisitive!

Acknowledgment

The authors would like to thank University Science Books for supporting this solutions manual, which we believe enhances the value of the associated text.

M. B. S.
E. V. A.
D. A. D.

CHAPTER

1

Introduction to Structure and Models of Bonding

S O L U T I O N S   T O   E X E R C I S E S

1.A.   B. 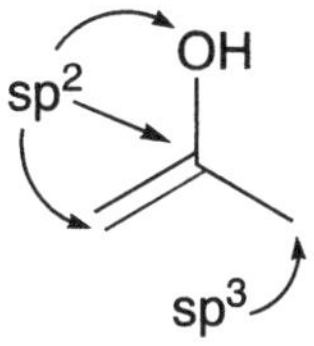   C. 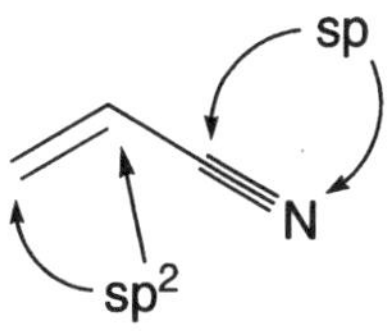   D. 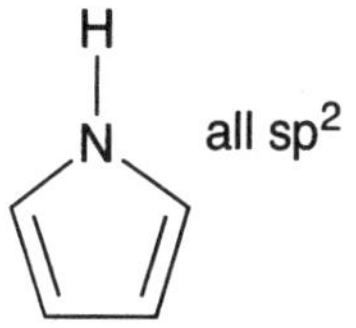   E.

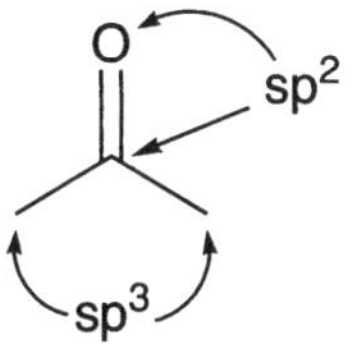

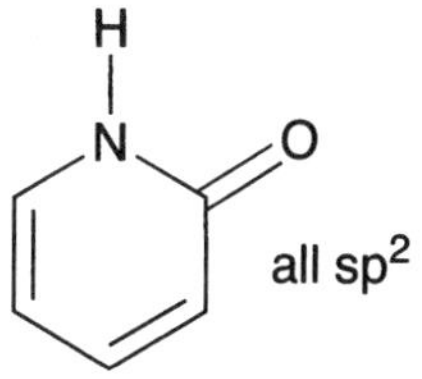

Note that the O in part B is better thought of as sp^2 rather than sp^3, so that the lone pair in the p orbital may best participate in resonance with the double bond.

2. Formal charges exist in A (+ on N)

                         D (+ on O), and

                         E (- on S).

Bond polarizations:

    B.   C.

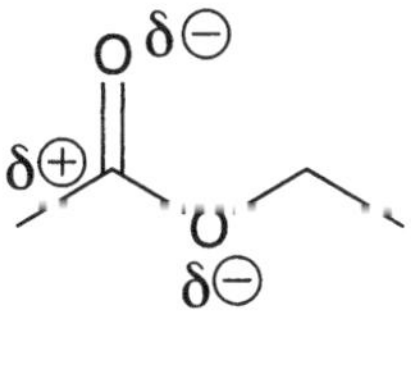

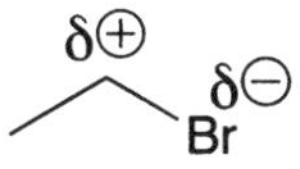

3.   B. 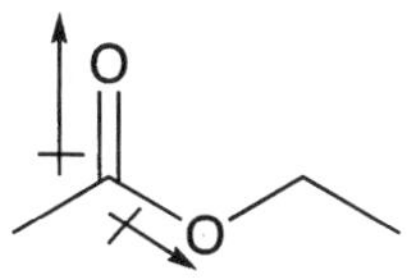   C.

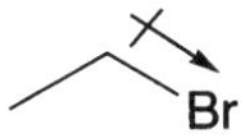

4. A.

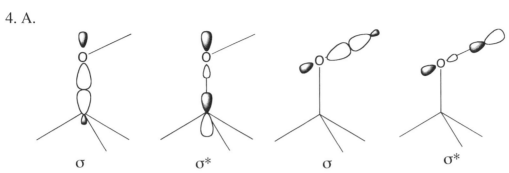

$$\sigma \qquad \sigma^* \qquad \sigma \qquad \sigma^*$$

Note that the polarization towards O is not obvious in the σ orbitals, since O atomic orbitals are smaller than C atomic orbitals due to the electronegativity of O. The polarization toward C or H is readily apparent in the antibonding MOs. (See Figure 1.17.) Note also that the component orbitals in the antibonding MOs are directed away from the center of the bond. This occurs due to negative overlap and cancellation in the center region.

B.

$$\sigma \qquad \qquad \sigma^*$$

Note that the Cl valence orbitals are larger than those of C are, so the polarization is most apparent in the σ orbital.

C.

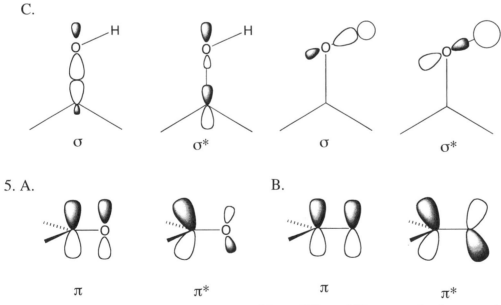

$$\sigma \qquad \sigma^* \qquad \sigma \qquad \sigma^*$$

5. A.   B.

$$\pi \qquad \pi^* \qquad \pi \qquad \pi^*$$

Note: CH_3 and H are nearly the same in electronegativity, so this π bond is nonpolar.

C.

π π^*

D.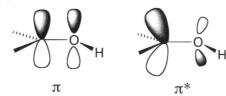

π π^*

Note: From the resonance structure that places the + charge on O, we might expect the opposite polarization than that in neutral acetone. However, the electronegativity differences are most important in determining the charge distribution. The charge is effectively delocalized to C (by resonance and induction) and to the three H's (by induction).

6.

1A.

1B.

1C.

1D.

1E.

5B. none

5C.

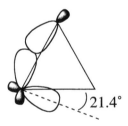

5D.

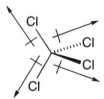

7. The small C-C-C angle of 60° implies that the orbitals involved in C-C bonding have a lot of p character. Eq. 1.1 gives $i = -2$, sp^{-2} hybridization – clearly an erroneous result. We should not expect the equation to work in this case, since the smallest bond angle that can be accommodated by s and p orbitals is 90°, corresponding to pure p orbitals. Therefore, the orbitals that make the C-C bonds in cyclopropane cannot be directed along the C-C axes, and the bonds must be "bent". The H-C-H angle of 118° is close to 120°, suggesting that the C orbitals involved in C-H bonding are approximately sp^2. Eq. 1.1 gives a hybridization of sp$^{2.13}$ for these orbitals. Since sp$^{2.13}$ hybrids are [1/(1+2.13)] x 100% = 31.9% s in character, the two C-H bonding hybrids will use up 63.9% of the s orbital, leaving 36.1% to be mixed into the C-C bonding hybrids (18.1% in each). So [1/(1+i)] x 100% = 18.1%, giving $i = 4.54$ or sp$^{4.54}$ hybridization. Eq. 1.1 gives an angle between these hybrids of arc cos(-1/4.54) = 102.7°, implying that the hybrids are misaligned with the C-C bond axes by (102.7°-60°)/2 = 21.4°.

8. The four bond dipoles are arranged symmetrically, such that the molecule has neither a net dipole moment nor a net quadrupole moment. Octapole moments are difficult to visualize and not often considered, but this arrangement of bond dipoles is analogous to an *f* orbital that has positive lobes pointing to alternate corners of a cube, *i.e.*, in tetrahedral directions.

9. In formaldehyde, the dipole from the charged resonance structure reinforces the inductive dipole arising from the higher electronegativity of O relative to C. In carbon monoxide, the dipole from the charged resonance structure opposes the inductive dipole, leading to cancellation. Therefore, the dipole moment of formaldehyde is larger. In addition, the shorter bond in carbon monoxide further reduces its dipole moment.

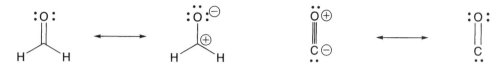

10. The C=O and H-O bond dipoles in conformer A are nearly parallel, leading to reinforcement and a higher dipole moment. Experimental values for A and B are 3.79 and 1.42 D, respectively.

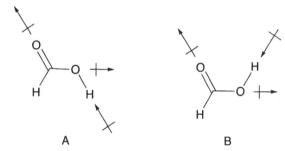

11. HF $\% \ ionic = 100\% * (1 - e^{-(4.0-2.1)^2/4}) = 59\%$

     HCl $\% \ ionic = 100\% * (1 - e^{-(3.0-2.1)^2/4}) = 18\%$

     HBr $\% \ ionic = 100\% * (1 - e^{-(2.8-2.1)^2/4}) = 12\%$

     HI $\% \ ionic = 100\% * (1 - e^{-(2.5-2.1)^2/4}) = 4\%$

     KF $\% \ ionic = 100\% * (1 - e^{-(4.0-0.8)^2/4}) = 92\%$

12. Several factors could be important. The zwitterionic resonance structure should be more important in formamide, tending to give it a higher dipole moment. However, formamide has an additional H–N bond dipole that leads to some cancellation, as noted for isomer B of formic acid in exercise 10. Experimentally, the two dipole moments are very similar: 3.79 D for formic acid and 3.73 D for formamide.

13. Group electronegativities depend most strongly on the nature (identity and hybridization) of the attached atom, while the rest of the group can play a stronger role in determining the dipole moment. The attached atoms of –C≡CH and –C≡N are very similar, leading to the same electronegativity. However, the C≡C bond is essentially nonpolar, while the C≡N bond is highly polar, leading to very different dipole moments.

14.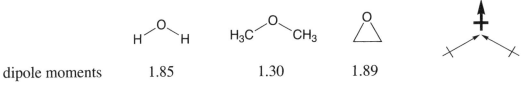

Note that none of these examples has a charge distribution shaped exactly like a d_{xy} orbital, but the symmetry of the charge distribution in each case is similar to the symmetry of a d orbital.

15. We can use Eq. 1.1 to calculate the C hybridization in the C-H bonds of CH_3Cl (H-C-H angle = 110.5°):

$$1 + i\cos(110.5°) = 0$$

$$i = 2.86$$

So the C hybridization in the C-H bonds of methyl chloride is $sp^{2.86}$.

For ethylene (H-C-H angle = 117.3°):

$$1 + i\cos(117.3°) = 0$$

$$i = 2.18$$

So the C hybridization in the C-H bonds of ethylene is $sp^{2.18}$.

16. In all three compounds, the dipole moments represent the vector sums of two H-O or C-O bond dipoles.

| dipole moments | 1.85 | 1.30 | 1.89 |

We will first compare the two ethers. In oxirane, the C-O-C angle is approximately 65° (a little larger than 60° due to the smaller covalent radius of O relative to C), leading to much less cancellation of the bond dipoles than for dimethyl ether, which has an angle of 111°. Therefore, oxirane has the higher dipole moment. If we want to be more quantitative, we can use simple trigonometry to show that the vector sum (and therefore dipole moment) should scale with $\cos(\theta/2)$, where θ is the bond angle. So if we wish to "predict" what the dipole moment of oxirane should be based on the value for dimethyl ether, we can set up a proportion:

$$\frac{\text{predicted dipole moment (oxirane)}}{1.30} = \frac{\cos(65°/2)}{\cos(111°/2)}$$

$$\text{predicted dipole moment (oxirane)} = 1.94$$

So the prediction based on angle is pretty close to the observed value. Note that the three-membered ring will also have three smaller effects on dipole moment. First, the greater p character of the C orbital in the C-O bond (see exercise 7) reduces the effective electronegativity of C, increasing the bond dipole (though a similar effect for O will partially counteract this). The other two effects both arise from the bent-bond nature of the C-O bonds. The reduced overlap of the C and O atomic orbitals increases the polarization of the bond, increasing the bond dipole. Also, the bending shortens the bond slightly (by about 2%), which directly lowers the bond dipole (Eq. 1.2). These effects are smaller than the angle effect and cancel each other out to some extent.

In comparing water and dimethyl ether, we must consider the substitution of CH_3 for H, but let's first consider the angle effect. Since CH_3 is larger than H, the C-O-C angle in dimethyl ether is expected to be larger than the H-O-H angle in water. The actual angles given in the text are 111° for dimethyl ether and 104.5° for water. This leads to a greater cancellation of the bond dipoles in dimethyl ether and a higher expected dipole moment for water:

$$\frac{\text{predicted dipole moment } (H_2O)}{1.30} = \frac{\cos(104.5°/2)}{\cos(111°/2)}$$

predicted dipole moment $(H_2O) = 1.41$

Thus, the angle effect does not come close to explaining the high dipole moment of water. The electronegativities of H and CH_3 (Tables 1.1 and 1.2) are 2.1 and 2.3. Using these values and 3.5 for O, the ionic characters of the H-O and C-O bonds (calculated as in exercise 11) are 39 and 30%, respectively. The magnitudes of the bond dipoles depend also on bond lengths (Eq. 1.2), and the H-O and C-O bond lengths are approximately 0.96 and 1.43 Å, respectively (Table 1.4). Therefore the relative length and ionic character effects on the H-O and C-O bond dipoles combine to give (0.39×0.96 Å)/(0.30×1.43 Å) = 0.87. This modifies our predicted value: 1.41×0.87 = 1.23. So our prediction is that the dipole moment of water should be even less than that of dimethyl ether! This is probably best explained by noting that electronegativity values are really averages and that the relative polarities of the H-O and C-O bonds in these compounds are not necessarily well modeled by this treatment. Indeed, the existence of multiple electronegativity scales that contain some large differences (Table 1.1) highlights this ambiguity. (Note that if we use the more approximate electronegativity of 2.5 for C, rather than 2.3 for CH_3, our predicted dipole moment for water comes out much closer at 1.68 D.)

GOING DEEPER

One might be tempted to explain the higher dipole moment for water relative to dimethyl ether as a consequence of hydrogen bonding. This cannot be correct, since the numbers given are for isolated molecules in the gas phase. Hydrogen bonding certainly can have a major impact on dipole moments in condensed phases, since it serves to further polarize and lengthen the O–H bonds. Experimental and theoretical studies on the dipole moment of a water molecule in liquid water or in ice provide values in the neighborhood of 3 D!

17. If we first consider the equatorial C–H bonds of cyclohexane, we find great similarity with benzene: the six bonds are roughly planar with bond dipoles that point towards the center of the ring. We should note, however, that these bond dipoles are much weaker than in benzene, due to the lower electronegativity of sp^3 carbon. The axial C–H bond dipoles also point into the ring, meaning that the entire exterior of cyclohexane has a slight positive charge, contrasting with the negative regions above and below the ring plane in benzene. Though small, the quadrupole moment of cyclohexane is expected to be non-zero, as it should be for any oblate-shaped molecule.

18. The strongest and most easily explained trend is the C–X bond length trend. As the halide gets bigger, the C–X bond length grows. As the C–X bond lengthens, the electrons in this bond move farther on average from the C atom, allowing the H–C–H angles to open slightly so as to reduce the interactions between the electrons involved in these bonds (a VSEPR argument). An increase in the H–C–H angle necessarily reduces the H–C–X angle. The electronegativity of X pulls electrons away from C, contributing to the same VSEPR effect. This argument on its own does not well explain the trend among the methyl halides, since the electronegativity decreases from F to I and the argument would predict a decreasing H–C–H angle. Note that this "contradiction" does not imply that the reasoning is faulty, but only that other effects are dominant.

Explaining the C–H bond lengths would be more difficult, since they do not follow a uniform trend. However, the fact that they were determined by "various methods" suggests that the small differences observed (≤ 0.02 Å) may not be meaningful.

G O I N G   D E E P E R

One should be cautious in comparing bond lengths from different methods. The issue is not just one of experimental error, but the very definition of bond length. For example, while electron and neutron diffraction measure distances between nuclei, X-ray diffraction measures distances between the centers of electron density. In molecules, atomic nuclei are not positioned exactly at the centers of electron density. These discrepancies can be especially large (0.06 Å or more) for bonds involving H, since the bonding electrons, the only ones reflecting the H position, can be significantly pulled toward the other atom. In microwave spectroscopy, atom positions are determined by matching calculated and observed rotational transitions, requiring different bond lengths definitions than the other methods. The differences between the methods in terms of bond angles are generally smaller.

Similar issues arise when comparing calculated and experimental bond lengths. Many molecular calculations do not account for zero-point vibrations, giving equilibrium bond lengths that correspond to the bottom of the potential well. Real molecules vibrate even in their ground state, and the anharmonicity of stretching vibrations leads to average bond lengths that are slightly longer than the equilibrium values.

19. These bond angles show that as F atoms replace H atoms on methyl radical, the geometry goes from nearly planar to pyramidal. Since F is highly electronegative, it bonds better with an sp^3-hybridized carbon, which is less electronegative than an sp^2-hybridized carbon. Therefore, addition of F atoms shifts the C hybridization from sp^2 to sp^3.

20. The individual VB C–H bond orbitals can be obtained through the following combinations of the methyl group orbitals:

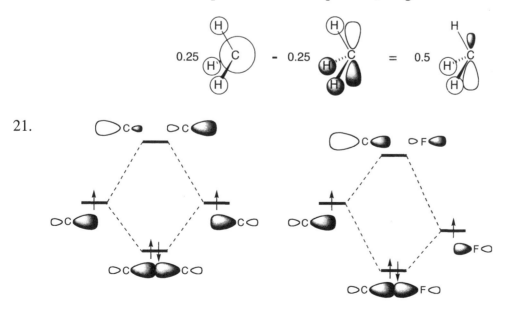

To understand these equations, remember that orbitals are just mathematical functions. If we add two orbitals that have the same sign (shading) around a given atom, then the resulting orbital will be reinforced in this region. If the signs are opposite, a cancellation will occur. It also helps to realize that subtraction is equivalent to reversing the shading pattern of the orbital to be subtracted and then adding. When three orbitals are involved, take just two at a time. As an intermediate step in the last two equations, we get:

21.

A few things should be noted about these diagrams: (1) The hybrid atomic orbital on F is both lower in energy and smaller than the one on C. (2) Both σ and σ* orbitals are lower in energy for the C-F bond. (3) The extent of mixing is less for the second-order mixing in the C-F bond. In other words, the non-zero energy gap between the C and F atomic hybrids results in a lower magnitude energy drop for the σ orbital and a lower energy increase for the σ* orbital. In spite of the reduced mixing, the bond strength of the C-F bond is greater, due to the larger stabilization of the electron that originates on C. (4) For both C-C and C-F, the σ* orbital is elevated more than the σ orbital is lowered. (5) Both σ and σ* C-F orbitals are polarized, though the polarization in the bonding orbital is only apparent through comparison to the sizes of the atomic orbitals.

The C-F bond reacts more readily with nucleophiles, since the unoccupied σ* orbital, which accepts the electrons from the nucleophile, is lower in energy. Nucleophiles attack the C-F bond at C, since the σ* orbital is polarized towards C.

22. To rationalize the polarization in the π MOs, we can construct them from their component π bonds:

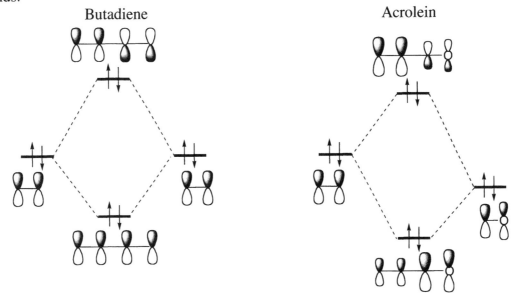

Butadiene   Acrolein

The butadiene π MOs are not polarized, while the acrolein higher π MO is polarized toward the C=C bond.

23.   Propene   Acetaldehyde

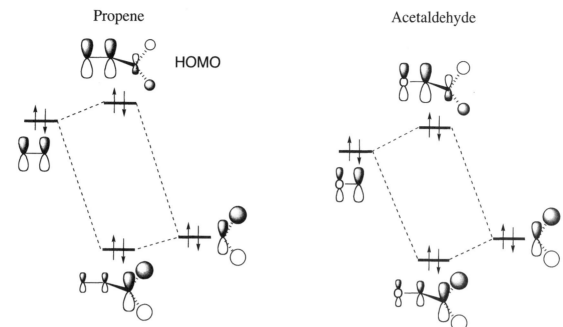

Since the π bonding orbital of C=O is lower in energy than that of C=C, the mixing of the C=O π orbital with π(CH₃) in acetaldehyde occurs with a smaller initial energy gap than the analogous mixing with C=C in propene. Therefore, the mixing is stronger in acetaldehyde, resulting in greater π(CH₃) character in the HOMO. The phase relationships are the same for the orbitals of the two molecules. (The polarization of the C=O π orbital toward O serves to reduce the mixing with π(CH₃), but this effect is less important.)

24. The π* orbital of propene can be constructed similarly to the π orbital (Figure 1.18 and exercise 23), realizing that the C=C π* group orbital will interact most with π*(CH₃):

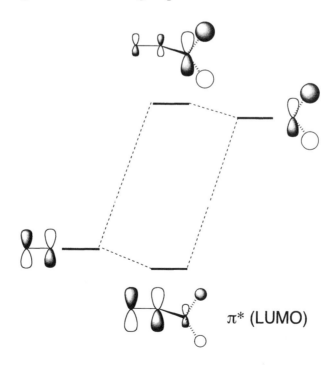

25. Comparing NH_3 and PH_3 with respect to VSEPR ideas, it is clear that the valence (bonding and nonbonding) electron pairs are closer to each other in NH_3 and will therefore more strongly repel each other. These repulsions are minimized with a roughly tetrahedral arrangement. In PH_3, where the repulsions are weaker, the differences between the lone and bonding electron pairs are accentuated. The lone pair is able to take up more space in this environment without causing excessive repulsions between the bonding pairs.

Considering hybridization and electronegativity effects, a couple points can be made. While a $2sp^3$ hybrid of N can very effectively overlap with an H 1s orbital, the overlap is smaller for a larger and more diffuse $3sp^3$ hybrid of P. In this case, overlap can be increased by using the unhybridized and more directional 3p orbital for the P–H bonds, giving H–P–H angles of approximately 90°. The lower electronegativity of P also means that the P–H bonds are nonpolar. Therefore, the bonding electrons are less concentrated on the central atom, leading to reduced repulsions between the bonding electrons (a point that further strengthens the VSEPR argument).

For an MO-based argument, see the chapter highlight in Section 1.4.2, "Pyramidal Inversion, NH_3 vs. PH_3."

26.

27.

π	π/lone pair	lone pair	π*

28. From the Appendix 3 figure, the HOMO of methylamine clearly arises from an out-of-phase mixing of the N p orbital and the π(CH₃) group orbital:

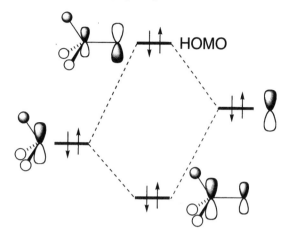

29. The interaction that occurs between the two sets of allyl π orbitals is just the single p-p interaction at the center of hexatriene. Realizing this, the orbitals of hexatriene are easily obtained:

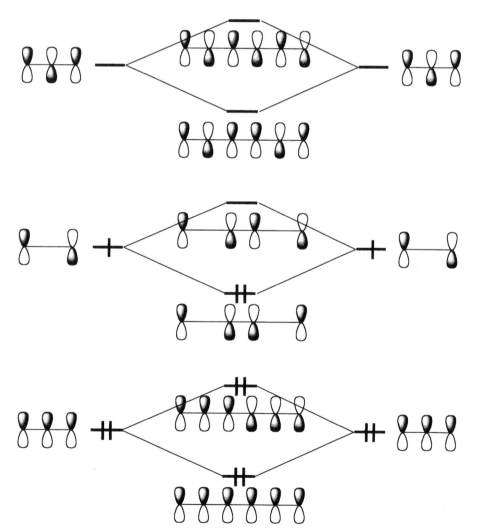

A few notes about this diagram:
1) The mixing shown above produces MOs with appropriate symmetry. If, for example, we tried to mix the bonding orbital of one allyl group with the nonbonding orbital of the other allyl group, the MOs obtained would not reflect the symmetry present in hexatriene. Therefore, this is not allowed (making it much easier for us to figure out what to do).
2) The energy of mixing in each case is smaller than the energy separating the allyl orbitals, since the mixing here results from one p-p interaction while the allyl orbitals result from two.
3) The HOMO and LUMO of hexatriene shown here are different from the MOs usually shown for hexatriene in that they have nodes on two carbon atoms. While these nodes are imposed by symmetry in allyl, they are not imposed by the symmetry of hexatriene. Through mixing among the MOs of hexatriene, these nodes can be shifted off the carbon atoms. Thus, we can approximate the hexatriene MOs by starting with two allyl groups, but we cannot get them exactly right.

30. The MH group orbitals come from the 2s and 2p orbitals of M and 1s orbital of H. The H 1s can mix only with the 2s and one of the 2p orbitals of M.

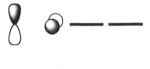

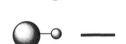

31. We can combine the group orbitals for CH_2 with those for MH in exercise 30 to get MOs for $H_2C=OH^+$:

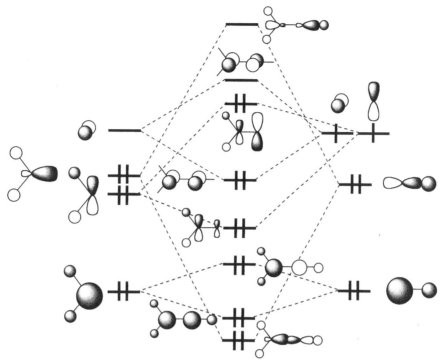

A few notes should be made about this diagram. First, the geometry used has the highest possible symmetry, that is with a 180° angle at O. Moving the H off the C–O axis could most easily be done by adjusting the OH group orbitals first in a Walsh diagram. Second, while relative energies may not be accurate, the general bonding scheme is consistent with expectations: one σ and one π bond. The in-plane π interaction is a filled-filled interaction.

32. The difference between allyl anion and an enolate ion is the substitution of O for C. In MO terms, this results in a lower energy p orbital on the O end, which leads to polarization of the MOs. (As in earlier exercises, polarization is more apparent in the higher energy MOs, since the O orbitals are contracted due to electronegativity.) The lack of symmetry of in the enolate ion leads to a non-zero coefficient on the central atom in the HOMO.

33. Taking the lone pair orbitals for formaldehyde from Figure 1.17 and then adding and subtracting, we get:

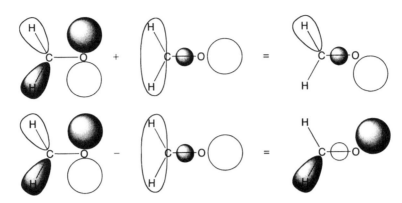

Note that these orbitals have O lobes aimed roughly in the direction of sp² hybrid orbitals, but that they also have C–H and C–C bonding character, as do the original MOs.

34. For CH₃, we started with seven AOs, so we must end up with seven MOs. Therefore, two MOs are not represented in Figures 1.7 and 1.8. These are $\pi^*(CH_3)$ orbitals:

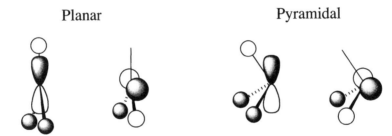

For CH₂, we started with six AOs, so we must end up with six MOs. Therefore, one MO is not represented in Figure 1.9: a $\pi^*(CH_2)$ orbital:

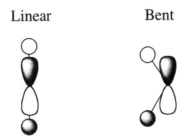

35. A trigonal bipyramidal geometry can be attributed to sp³d hybridization, but the five hybrid orbitals are not equivalent. The equatorial hybrids are sp², and the apical hybrids are pd:

Electronegative ligands will prefer the pd hybrids, since they have no s character. Electrons in these hybrids will be more available to the ligands.

2

Strain and Stability

Note: All energy values are in kcal/mol.

1. The twist-boat conformation has a higher enthalpy than the chair conformation by 5.5 kcal/mol (Figure 2.12). If we presume the entropies are the same (including an expected statistical factor that would favor the twist-boat due to its greater number of conformations), then $\Delta G = 5.5$ kcal/mol. We can then compute the equilibrium constant at 298K:

$$K = e^{-\Delta G/RT} = e^{-(5.5 \text{ kcal/mol})/(0.00199 \text{ kcal/mol K})(298\text{K})} = 9.4 \times 10^{-5}$$

So the percentage of the twist-boat conformation is $9.4 \times 10^{-3}\% = 0.0094\%$.

2.

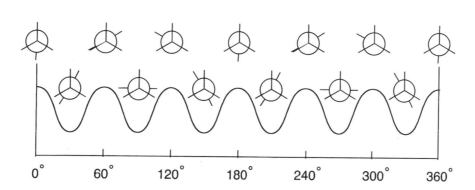

3. The central C–C–C angle is 128° (Fig. 2.21). Using Eq. 1.1:

$i = 1.62$

Therefore, the central C hybrids that bond to the t-butyl groups appear to be $sp^{1.62}$ (assuming that the bonds are not bent like those in cyclopropane – see exercise 7 in Chapter 1). In contrast, the corresponding hybrids in propane are very close to sp^3.

G O I N G   D E E P E R

We can check further whether the hybridization makes sense by calculating what the hybridization must be for the orbitals that bond to H (using the same method as in exercise 7 in Chapter 1):

2(s character in C-C) + 2(s character in C-H) = 1

$$2\left(\frac{1}{1+1.62}\right) + 2\left(\frac{1}{1+x}\right) = 1$$

$$\frac{1}{1+x} = 0.118$$

$$x = 7.45$$

The C–H bonds apparently must use nearly pure p orbitals ($sp^{7.45}$), giving a predicted H–C–H angle of 98°. Unfortunately, we do not have this angle to compare to the calculation, but we can say that if the H–C–H angle is not compressed all the way to 98°, then the C–C the bonds are apparently bent due to strain. Molecular mechanics calculations put the H–C–H angle near 104°. Redoing the calculation in the other direction with this as a starting point gives $sp^{4.1}$ hybridization for the C–H bonds and $sp^{2.3}$ hybridization for the orbitals involved in the C–C bonds. The orbitals would be separated by 116°. This suggests that the C–C bonds are each bent by (128°-116°)/2 = 6°. Thus, strain and bent bonds are not limited to cyclic compounds!

4. The CH group from 1,3-butadiene would correspond to the $C_d–(C_d)(H)$ group increment of 6.78 kcal/mol. We can first verify that this group increment is representative of butadiene:

$$\Delta H_f^\circ(1,3\text{-butadiene}) = 2\Delta H_f^\circ[C_d–(H)_2] + 2\Delta H_f^\circ[C_d–(C_d)(H)]$$

$$= 2(6.26) + 2(6.78)$$

$$= 26.08$$

This compares very well with the experimental ΔH_f° for butadiene. We can now calculate the "aromaticity-free" ΔH_f° for benzene:

$$\Delta H_f^\circ(\text{benzene}) = 6\Delta H_f^\circ[C_d–(C_d)(H)] = 6(6.78) = 40.68$$

Since the actual ΔH_f° for benzene is 19.7 kcal/mol, the stabilization due to aromaticity is 21.0 kcal/mol. This value is considerably lower than the estimate of 32 kcal/mol computed from the isodesmic reaction of cyclohexene. We can obtain the latter value also through group increments, using the CH from cyclohexene:

$$\Delta H_f^\circ(\text{benzene}) = 6\Delta H_f^\circ[C_d-(H)(C)] = 6(8.59) = 51.54$$

Comparing to the experimental value gives an aromatic stabilization of 31.8 kcal/mol.

The value that is more appropriate depends on one's intentions. If one is interested in aromatic stabilization that goes beyond stabilization due to the conjugation in an acyclic model, the value of 21 kcal/mol is more appropriate. If one is interested in the entire stabilization due to resonance, the value of 32 kcal/mol is more appropriate.

5. A shorter than normal C=C bond is not generally what one would expect for a double bond that is twisted and therefore weaker. Any misalignment and pyramidalization of the "p" orbitals of the π bond should weaken the bond. Furthermore, pyramidalization will lead to more p character in the σ hybrids, and that would tend to lengthen, not shorten the bonds. If hybridization is the explanation, the shorter C=C bond implies that the sp^2 hybrids involved in the C=C bond are more "sp-like." This is indeed consistent with the stresses at the double bond, which will tend to widen the C-C-C angles, making the molecule flatter. The excess p character is presumably used in the hybrids involved in the C-H bonds, consistent with smaller H-C-C angles. (Note that the bond contraction is small, <0.02 Å, so the hybridization changes are presumably also small.)

⇓ stress on alkene C's due to ring constraint

⇑

6. The axial *t*-butyl group in the right-hand conformer has no 1,3-diaxial interactions with other atoms as it would have in the corresponding cyclohexane (with CH_2 groups replacing O). Lone pairs are held more closely to an atom than are bonded atoms, so their steric interactions are more severe in determining atomic hybridization (VSEPR) but less severe in interatomic interactions. The axial methyl in the left-hand conformer actually experiences larger 1,3-diaxial interactions than it would in the corresponding cyclohexane, because the two intervening bonds are C–O bonds, which are shorter than C C bonds.

7. By analogy to the isodesmic approach of Eq. 2.32, the following isodesmic reaction can be used for naphthalene:

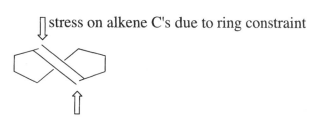

Balancing this equation by inspection can be difficult, but we can do it easily with a little algebra. We have already made the carbons balance, so we need only consider hydrogen:

$(n+m)(12H) = n(8H) + m(18H)$

$4n = 6m$

$n/m = 3/2$

So the balanced equation is:

$\Delta H°_{rxn} = \Sigma[\Delta H_f°(\text{products})] - \Sigma[\Delta H_f°(\text{reactants})]$

$= 3(36) + 2(-43.5) - 5(6.22)$

$= -10.1$

The meaning of this heat of reaction is less straightforward than that for Eq. 2.32. Let us separate the aromaticity into that from the first ring and that from the second ring. In this reaction, 2 molecules lose the first-ring aromaticity and 3 gain the second-ring aromaticity. We will assume that the first-ring aromaticity is the same as that calculated for benzene. We will also assume that naphthalene is unstrained. The strain energy of *trans*-decalin can be taken from Fig. 2.15. The strain energy of tetralin can be calculated through group increments:

strain E, tetralin = $\Delta H_f°(\text{exp.}) - \Delta H_f°(\text{strain-free})$

$= \Delta H_f°(\text{exp.}) - \{4\Delta H_f°[C_B–(H)] + 2\Delta H_f°[C_B–(C)] + 2\Delta H_f°[C–(C_B)(C)(H)_2] +$

$2\Delta H_f°[C–(H)_2(C)_2]\}$

$= 6.22 - \{4(3.30) + 2(5.51) + 2(-4.86) + 2(-4.93)\}$

$= 6.22 - 4.64 = 1.58$ (Note: similar to the strain energy of cyclohexene of 1.4)

Now we are prepared to analyze the heat of this isodesmic reaction:

$\Delta H°_{rxn} = 2(\text{first-ring arom.}) - 3(\text{second-ring arom.}) + 5(\text{strain E, tetralin}) - 2(\text{strain E, }t\text{-decalin})$

$-10.1 = 2(32) - 3(\text{second-ring arom.}) + 5(1.58) - 2(-1.9)$

second-ring. arom. = 29 kcal/mol

So the aromaticity of the second ring is slightly less than that of the first ring. Alternatively, we could use *cis*-decalin, but the difference in $\Delta H°_{rxn}$ and the strain energy correction would cancel out, giving the same result.

As shown in exercise 4, a roughly equivalent approach to the isodesmic reaction of Eq. 2.32 is to use the group increment for benzene: $\Delta H_f°[C_d–(H)(C)]$. Analogously for naphthalene:

energy due to aromaticity = $\Delta H_f°$(exp.) - $\Delta H_f°$(arom.-free)

$$= \Delta H_f°(\text{exp.}) - \{8\Delta H_f°[C_d–(H)(C)] + 2\Delta H_f°[C_d–(C)_2]\}$$

$$= 36 - \{8(8.59) + 2(10.34)\}$$

$$= 36 - 89.4 = -53 \text{ kcal/mol}$$

By this method the total aromaticity is 53 kcal/mol, so the second-ring aromaticity is 53 - 32 = 21 kcal/mol, somewhat less than the 29 calculated in the isodesmic reaction.

Taking the message of exercise 4, a more reasonable approach is to use all sp^2 group increments:

energy due to aromaticity = $\Delta H_f°$(exp.) - $\Delta H_f°$(arom.-free)

$$= \Delta H_f°(\text{exp.}) - \{8\Delta H_f°[C_d–(C_d)(H)] + 2\Delta H_f°[C_d–(C_d)_2]\}$$

$$= 36 - \{8(6.78) + 2(4.6)\}$$

$$= 36 - 63.4 = -27.4 \text{ kcal/mol}$$

The most appropriate first-ring aromaticity in this case is the similarly calculated value for benzene of 21 kcal/mol (see exercise 4). This gives a much smaller second-ring aromaticity of 27.4 - 21.0 = 6.4 kcal/mol.

In summary, the second ring values from the three calculations are all smaller than the first-ring values, though the amount varies from 91% to 66% to 30%. As noted in exercise 4, the most appropriate method and result depends on how the number will be used.

8. The strain-free group increment calculation is the same for all four isomers:

$$\Delta H_f° = 4\Delta H_f°[C–(H)(C)_3] + 6\Delta H_f°[C–(H)_2(C)_2]$$

$$= 4(-1.90) + 6(-4.93)$$

$$= -37.18$$

strain E = -32.12 - (-37.18) = 5.06 kcal/mol

strain E = -14.38 - (-37.18) = 22.80 kcal/mol

strain E = -24.46 - (-37.18) = 12.72 kcal/mol

strain E = -20.54 - (-37.18) = 16.64 kcal/mol

Adamantane does have the lowest strain energy and is nearly strain-free. These results are consistent with it being the $C_{10}H_{16}$ thermodynamic sink.

9. In the equatorial conformer, considerable allylic strain is present, seen more clearly from "overhead":

10. We normally expect more highly substituted alkenes to be more stable, but in this case the more highly substituted isomer, B, suffers from substantial allylic strain. Conjugation tends to keep the pyrrolidine ring roughly planar with the cyclohexane ring, which can lead to an unfavorable A-strain. In isomer A, the conjugation can be accommodated by placing the methyl in the axial position:

11. As for hydrogen peroxide (see text), there are competing effects. The best alignment of a good donor (lone pair) with a good acceptor (N-H) occurs in the gauche conformation, while the number of gauche interactions is minimized in the anti conformation. The actual preferred conformation is a compromise, having a dihedral angle of 90-95° between the lone pairs.

gauche   anti   actual

12. The strain-free ΔH_f° values are identical for the corresponding members of series A and B:

A   **B**

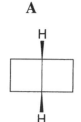

$\Delta H_f^\circ(\text{strain-free}) = 2\Delta H_f^\circ[\text{C–(H)(C)}_3] + 4\Delta H_f^\circ[\text{C–(H)}_2(\text{C})_2]$

$= 2(-1.90) + 4(-4.93) = -23.52$

Strain energy = 25.63 - (-23.52)   Strain energy = 16.37 - (-23.52)

= 49.15   = 39.89

Additivity of rings (4 + 4)   Additivity of rings (4 + 5)

expected strain = 2(27.6) = 55.2   expected strain = 27.6 + 6.3 = 33.9

(Note: Any bicyclic ring system has three identifiable rings. The ring additivity method uses the smaller two. The series A example above has two 4-membered and an 8-membered ring. The series B example has a 4-membered and two 5-membered rings.)

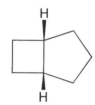

$\Delta H_f^\circ(\text{strain-free}) = 2\Delta H_f^\circ[\text{C–(H)(C)}_3] + 5\Delta H_f^\circ[\text{C–(H)}_2(\text{C})_2]$

$= 2(-1.90) + 5(-4.93) = -28.45$

Strain energy = 0.51 - (-28.45)   Strain energy = -12.42 - (-28.45)

= 28.96   = 16.03

Additivity of rings (4 + 5)   Additivity of rings (5 + 5)

expected strain = 27.6 + 6.3 = 33.9   expected strain = 2(6.3) = 12.6

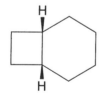

$\Delta H_f^\circ(\text{strain-free}) = 2\Delta H_f^\circ[\text{C–(H)(C)}_3] + 6\Delta H_f^\circ[\text{C–(H)}_2(\text{C})_2]$

$= 2(-1.90) + 6(-4.93) = -33.38$

Strain energy = -6.1 - (-33.38)   Strain energy = -23.04 - (-33.38)

= 27.28   = 10.34

Additivity of rings (4 + 6)   Additivity of rings (5 + 6)

expected strain = 27.6 + 0 = 27.6   expected strain = 6.3 + 0 = 6.3

The ring additivity method gives reasonably good values in all cases.

13. Most tetraalkylethanes show a preference for the gauche conformer due to geminal repulsions that cause an increase in the geminal C-C-C angles. In bicyclopentyl, the 5-membered rings prevent this increase in the geminal angles. The anti conformation is therefore preferred, minimizing the number of gauche interactions.

14. We can calculate ΔH_f° from group increments both ways:

Aromatic: $\Delta H_f^\circ = 60(\Delta H_f^\circ[C_{BF}-(C_{BF})_3]) = 60(1.5) = 90$

Double bonds: $\Delta H_f^\circ = 60(\Delta H_f^\circ[C_d-(C_d)_2]) = 60(4.6) = 276$

Clearly, the second value is closer to the experimental value of 634.8 kcal/mol, so it appears that C_{60} is better thought of as a collection of double bonds. From this viewpoint, the molecule has a strain energy of about 360 kcal/mol. On a per C basis, the strain of C_{60} (6 kcal/mol) is therefore similar to that of cyclobutane (6.6 kcal/mol). If treated as aromatic, the strain energy per C would be 9 kcal/mol, the same as that for that for cyclopropane. Without data concerning expected strain values for similarly distorted sp^2 C's, this is a difficult judgment call.

An intermediate view has also been proposed for C_{60}: a molecule with aromatic six-membered rings and non-aromatic five-membered rings.

15. C_2H_4 C_2H_6

$\Delta H_f^\circ = 2\Delta H_f^\circ[C_d-(H)_2]$ $\Delta H_f^\circ = 2\Delta H_f^\circ[C-(H)_3(C)]$

$= 2(6.76) = 12.52$ $= 2(-10.20) = -20.40$

Strain energy = 12.5 - 12.52 = 0.0   Strain energy = -20.02 - (-20.40) = 0.38

Olefin strain = 0.0 - 0.38 = -0.4 kcal/mol

From this result, one could call C_2H_4 a hyperstable olefin, but it is more appropriate to say that both C_2H_4 and C_2H_6 are unstrained. The small difference is within the expected error of the group increment values (and perhaps also the experimental values).

$\Delta H_f^\circ = 2\Delta H_f^\circ[C_d-(H)(C)] +$ $\Delta H_f^\circ = 6\Delta H_f^\circ[C-(H)_2(C)_2]$

$2\Delta H_f^\circ[C-(C_d)(C)(H)] +$ $= 6(-4.93) = -29.58$

$2\Delta H_f^\circ[C-(H)_2(C)_2]$ Strain energy = -29.93 - (-29.58) = -0.35

$= 2(8.59) + 2(-4.76) + 2(-4.93) = -2.20$

Strain energy = -1.20 - (-2.20) = 1.00

Olefin strain = 1.0 - (-0.35) = 1.35 kcal/mol

$\Delta H_f^\circ = \Delta H_f^\circ[C_d-(C)_2] + \Delta H_f^\circ[C_d-(H)(C)] +$

$\qquad 3\Delta H_f^\circ[C-(C_d)(C)(H)] +$

$\qquad \Delta H_f^\circ[C-(H)(C)_3] +$

$\qquad 3\Delta H_f^\circ[C-(H)_2(C)_2]$

$\qquad = 10.34 + 8.59 + 3(-4.76) + (-1.90) +$

$\qquad 3(-4.93) = -12.04$

Strain energy = 11.91 - (-12.04) = 23.95

(compare to Figure 2.15 value: 24)

$\Delta H_f^\circ = 2\Delta H_f^\circ[C-(H)(C)_3] +$

$\qquad 7\Delta H_f^\circ[C-(H)_2(C)_2]$

$\qquad = 2(-1.90) + 7(-4.93) = -38.31$

Strain energy = -23.26 - (-38.31) = 15.05

Olefin strain = 23.95 - 15.05 = 8.90 kcal/mol

$\Delta H_f^\circ = \Delta H_f^\circ[C_d-(C)_2] + \Delta H_f^\circ[C_d-(H)(C)] +$

$\qquad 3\Delta H_f^\circ[C-(C_d)(C)(H)] +$

$\qquad \Delta H_f^\circ[C-(H)(C)_3] +$

$\qquad 4\Delta H_f^\circ[C-(H)_2(C)_2]$

$\qquad = 10.34 + 8.59 + 3(-4.76) + (-1.90) +$

$\qquad 4(-4.93) = -16.97$

Strain energy = 3.34 - (-16.97) = 20.31

$\Delta H_f^\circ = 2\Delta H_f^\circ[C-(H)(C)_3] +$

$\qquad 8\Delta H_f^\circ[C-(H)_2(C)_2]$

$\qquad = 2(-1.90) + 8(-4.93) = -43.24$

Strain energy = -23.88 - (-43.24) = 19.36

Olefin strain = 20.31 - 19.36 = 0.95 kcal/mol

$$\Delta H_f^\circ = \Delta H_f^\circ[C_d-(C)_2] + \Delta H_f^\circ[C_d-(H)(C)] +$$

$$3\Delta H_f^\circ[C-(C_d)(C)(H)] +$$

$$\Delta H_f^\circ[C-(H)(C)_3] +$$

$$5\Delta H_f^\circ[C-(H)_2(C)_2]$$

$$= 10.34 + 8.59 + 3(-4.76) + (-1.90) +$$

$$5(-4.93) = -21.90$$

Strain energy = -8.95 - (-21.90) = 12.95

$$\Delta H_f^\circ = 2\Delta H_f^\circ[C-(H)(C)_3] +$$

$$9\Delta H_f^\circ[C-(H)_2(C)_2]$$

$$= 2(-1.90) + 9(-4.93) = -48.17$$

Strain energy = -26.43 - (-48.17) = 21.74

Olefin strain = 12.95 - 21.74 = -8.79 kcal/mol

This is clearly a case of a hyperstable olefin, even though it is a bridgehead alkene! But it makes sense, given that the trans alkene is contained in a 10-membered ring. The alkene presumably reduces adverse transannular and/or torsional interactions compared to the saturated system.

16.

A   **B**

If we presume that the methyl and isopropyl groups independently affect this equilibrium, then we can simply add the ΔH° and ΔS° values. Since the methyl goes from equatorial to axial, its contributions are as given in the exercise, while the signs of ΔH° and ΔS° must be reversed for isopropyl. Therefore,

$$\Delta H^\circ = 1.75 - 1.52 = 0.23 \text{ kcal/mol}$$

$$\Delta S^\circ = 0 - (-2.31) = 2.31 \text{ cal/mol K}$$

	300 K	100 K	75 K
$\Delta G^\circ\ (= \Delta H^\circ - T\Delta S^\circ)$	-0.46	0.0	0.06
$K\ (= e^{-\Delta G^\circ/RT})$	2.2	1.0	0.67
$\dfrac{B}{A} \left(= \dfrac{100*K/(K+1)}{100*1/(K+1)} \right)$	68/32	50/50	40/60

Note that whether **A** or **B** is the preferred conformation depends on temperature.

GOING DEEPER

Can we understand why $\Delta S°$ is negative for isopropyl and zero for methyl and why the magnitude of $\Delta H°$ is larger for methyl? The $\Delta S°$ value shows that isopropylcyclohexane has less entropy with the isopropyl group axial. When in the axial position, rotation about the bond between the cyclohexane and the isopropyl should be substantially hindered by 1,3-diaxial interactions. The isopropyl group will be less free to rotate, and since entropy reflects the degrees of freedom in the system, we see a reduction in entropy for axial isopropylcyclohexane. Rotation about the corresponding C-C bond in methylcyclohexane should be comparably unhindered in both the equatorial and axial conformers. We can explain the lower $\Delta H°$ for isopropylcyclohexane by noting that the equatorial conformer also suffers steric interactions, though not as bad as those in the axial conformer. In the equatorial conformer, rotation of the isopropyl group does not help to alleviate these interactions:

17. We first need to calculate the strain-free $\Delta H_f°$ values, those calculated from group increments without correction.

6 C_B–(C)	6(5.51)	4 C_d–(C)$_2$	4(10.34)	6 C–(C)$_4$	6(0.50)
6 C–(H)$_3$(C)	6(-10.20)	2 C–(C$_d$)$_2$(C)$_2$	2(2.86)	6 C–(H)$_3$(C)	6(-10.20)
	———	6 C–(H)$_3$(C)	6(-10.20)		———
	$\Delta H_f° = -28.14$		$\Delta H_f° = -14.12$		$\Delta H_f° = -58.2$

Two of the increments above do not appear in Table 2.4. Since no value is given for C–(H)$_3$(C$_B$), the value for C–(H)$_3$(C) was used instead. This can be justified by noting that values for C–(H)$_2$(C)$_2$ and C–(C$_B$)(C)(H)$_2$ are nearly the same. A value for C–(C$_d$)$_2$(C)$_2$ was obtained by extrapolating from the values for C–(C)$_4$ = 0.50 and C–(C$_d$)(C)$_3$ = 1.68; these differ by 1.18 kcal/mol, so C–(C$_d$)$_2$(C)$_2$ = 1.68 + 1.18 = 2.86 is a reasonable estimate.

The strain energy is equal to the experimental $\Delta H_f°$ minus the calculated strain-free $\Delta H_f°$.

experimental	$\Delta H_f° = -24.0$	$\Delta H_f° = 25.5$	$\Delta H_f° = 67.2$
strain-free	$\Delta H_f° = -28.14$	$\Delta H_f° = -14.12$	$\Delta H_f° = -58.2$
strain energy	4.1	39.6	125.4

18.

The three hydroxyl groups point in roughly the same direction, as shown. The hydroxyl groups could participate in hydrogen bonding with a multifunctional acceptor (like a carbohydrate) or serve to form a metal ion chelate. The structure shown clearly has a hydrophobic face (top) and a hydrophilic face (bottom), suggesting that it may act also as a surfactant. Given that cholic acid is a component of bile, which is important in emulsification of fats, this last suggestion seems likely to be the most important.

19. Linking the flagpole positions of boat cyclohexane with one carbon gives norbornane. Its strain energy (given in Figure 2.15) is 17.0 kcal/mol.

20. This is a case like FCH_2NH_2, discussed in the text, where a very good donor on one atom (lone pair) can interact with a very good acceptor on the next atom (C-F bond). Aligning these groups *antiperiplanar* gives a 90° F-C-O-H dihedral angle, if we consider sp^2 hybridization at O. The 90° angle leads to a nearly eclipsed arrangement of two H atoms, and so we can expect some adjustment. In the actual molecule, the F-C-O-H dihedral angle is ~70°. Note that an sp^3 hybridization model for oxygen would predict a 60° dihedral angle, also in acceptable agreement with experiment.

21. We need ΔH for the following reaction:

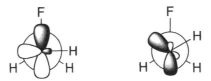

	$8\ CO_2$	+	$4\ N_2$
ΔH_f° ?	-94.05		0

We can calculate ΔH_f° for octanitrocubane using group increments, but Table 2.4 does not have increments for the NO_2 group. We were given ΔH_f° for CH_3NO_2, and we can use that to estimate a value for the NO_2 group increment. (We say "estimate" because the values in Table 2.4 come from experimental data on many compounds, not just one. Group increments for nitro-containing compounds do exist, and though these are set up somewhat differently than those that follow, the analysis below will serve our purposes. Realize that this computation can be done in different ways.)

$$\Delta H_f^{\circ}(CH_3NO_2) = \Delta H_f^{\circ}[C-(N)(H)_3] + \Delta H_f^{\circ}[NO_2-(C)]$$

$$-19.3 = -10.08 + \Delta H_f^{\circ}[NO_2-(C)]$$

$$\Delta H_f^{\circ}[NO_2-(C)] = -9.2$$

Now we can calculate ΔH_f° for octanitrocubane. We want a value that approximates the experimental value, so we must add in the strain energy, which we can get from Fig. 2.15, assuming it to be the same as the parent cubane.

$$\Delta H_f^{\circ}(octanitrocubane) = 8\Delta H_f^{\circ}[C-(N)(C)_3] + 8\Delta H_f^{\circ}[NO_2-(C)] + strain\ energy$$

$$= 8(-3.2) + 8(-9.2) + 166$$

$$= 66.8$$

$$\Delta H_{rxn}^{\circ} = \Sigma[\Delta H_f^{\circ}(products)] - \Sigma[\Delta H_f^{\circ}(reactants)]$$

$$= 8(-94.05) + 4(0) - (66.8)$$

$$= -819\ kcal/mol$$

So approximately 819 kcal/mol would be produced in the reaction. (This value is somewhat less than a recent literature estimate of 873 kcal/mol.)

22. The cal unit was originally defined as the energy required to raise the temperature of 1 g of water by 1°C. So the specific heat of water is 1 cal/g°C. Since the density of water is 1 g/mL, the heat capacity is 1 cal/mL°C = 1 kcal/L°C.

Heat required to increase temp. of 1 L of water by 25° = (1 L)(25°)(1 kcal/L°C)

$$= 25\ kcal$$

mol octanitrocubane = 25 kcal/(819 kcal/mol) = 0.031 mol

mass octanitrocubane = 0.031 mol × 464 g/mol = 14 g

For comparison, it would take about 16 g of nitroglycerin but quite a bit more TNT – as much as 48 g, assuming no external oxygen supply. The effectiveness of an explosive depends on a number of factors besides exothermicity, including oxygen balance, moles of gas produced, molecular weight of gases produced, and density. Octanitrocubane has perfect oxygen balance (no O_2 needed or produced) and compares favorably in the other categories with explosives in use. This energetic compound was first made in 2000.

23.(a) In going from a linear to a branched alkane, two CH_2 groups are replaced by a CH and a CH_3 group:

$$2\Delta H_f^\circ[C-(H)_2(C)_2] \qquad\qquad \Delta H_f^\circ[C-(H)(C)_3] + \Delta H_f^\circ[C-(H)_3(C)]$$

$$= 2(-4.93) \qquad\qquad\qquad = -1.90 + (-10.20)$$

$$= -9.86 \qquad\qquad\qquad\qquad = -12.10$$

The statement is true. The calculation seems to show that increasing the number of methyl groups is what makes branched alkanes more thermodynamically stable, in spite of the lower stability of the branched carbon itself. One must be careful in making such conclusions, however, since we're comparing groups that contain different numbers of hydrogen atoms.

(b) In going from a terminal to an internal alkene, a double-bonded CH_2 is replaced by an double-bonded CH, and an sp^3 CH_2 is replaced with a CH_3:

$$\Delta H_f^\circ[C_d-(H)_2] + \Delta H_f^\circ[C-(C_d)(C)(H)_2] \qquad \Delta H_f^\circ[C_d-(H)(C)] + \Delta H_f^\circ[C-(C_d)(H)_3]$$

$$= 6.26 + (-4.76) \qquad\qquad\qquad\qquad = 8.59 + (-10.03)$$

$$= 1.50 \qquad\qquad\qquad\qquad\qquad\qquad = -1.44$$

(To do the right-hand calculation, we needed a group increment that is not in Table 2.4: $[C-(C_d)(H)_3]$. It was estimated by adjusting the value for $[C-(H)_3(C)]$ by the difference between $[C-(H)_2(C)_2]$ and $[C-(C_d)(C)(H)_2]$.)

The internal double bond is indeed more stable.

(c) Consider the two reactions separately:

$$\Delta H^\circ_{rxn} = \Sigma[\Delta H_f^\circ(\text{prod.})] - \Sigma[\Delta H_f^\circ(\text{react.})] \text{ (Note: only the groups that change need be included)}$$

$$= (\Delta H_f^\circ[C-(H)_2(C)_2] + \Delta H_f^\circ[C-(H)_3(C)]) - (\Delta H_f^\circ[C_d-(H)(C)] + \Delta H_f^\circ[C_d-(H)_2] + \Delta H_f^\circ(H_2))$$

$$= (-4.93 + (-10.20)) - (8.59 + 6.26 + 0)$$

$$= -29.98$$

$$\Delta H°_{rxn} = (\Delta H_f°[C–(O)(C)(H)_2] + \Delta H_f°[O–(C)(H)]) - (\Delta H_f°[CO–(C)(H)] + \Delta H_f°(H_2))$$

$$= (-8.1 + (-37.9)) - (-29.1 + 0)$$

$$= -16.9$$

Hydrogenation of olefins does appear to be more exothermic.

24. The geminal phenyl groups in 1,1,2,2-tetraphenylethane can line up approximately face-to-face, relieving the geminal repulsion. Therefore, the anti conformation, having fewer gauche interactions, is preferred.

25. It appears that in their lowest energy conformations with respect to rotation about the C–Ph or C–*i*Pr bonds, the axial Ph group would suffer fewer steric interactions than would the axial *i*Pr group (due to the C–H aimed between the two axial H's). However, the *i*Pr group can rotate through a larger angle without much increase in the steric interactions, while the Ph group has less freedom to rotate out of its lowest energy position. So the Ph group must pay a higher entropic price in order to keep the enthalpic costs lower. Since the A values are free energy quantities, both enthalpy and entropy are important. (We are *not* saying that the rotations around these bonds are slow.)

26. The envelope conformation of cyclopentane has two sets of eclipsed C-H bonds, while cyclopentene has only one (with the eclipsed C-H bonds farther apart, due to the larger C–C–H angles at the sp^2 alkene C's).

27. Consider the contributions of the quadratic and cubic terms to the shape of the potential function separately (taking $k_r' = -k_r$):

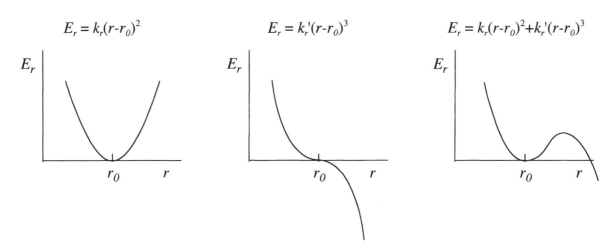

$$E_r = k_r(r-r_0)^2 \qquad\qquad E_r = k_r'(r-r_0)^3 \qquad\qquad E_r = k_r(r-r_0)^2+k_r'(r-r_0)^3$$

The function on the right resembles the Morse potential in the region where r is near to or less than r_0. However, for $r \gg r_0$, the energy drops instead of staying level. This is the long bond catastrophe. A molecular mechanics program using this equation to calculate a structure with a bond that is too long will tend to make the bond even longer. One way to have the program avoid this problem is to check the slope of the energy curve. If the slope is negative and $r > r_0$, then the bond is in this "long bond" regime and should be made shorter.

28. Dipole moments can be calculated by summation of bond dipoles, which are in turn readily calculated from atomic electronegativities. Such calculations require nothing more than the molecular geometries. However, polarizability has to do with electron clouds and the ease of their deformation, and molecular mechanics calculations have nothing at all to do with electrons or orbitals. This does not mean that a force field could not be parameterized to give reasonable polarizabilities, but the method is more naturally suited to handle dipole moments.

29. Spirononane is essentially unstrained, while [5.3.5.3]fenestrane is a very highly strained molecule. The accuracy of molecular mechanics is generally poor for strained molecules, since molecules with similar distortions are not well represented in the parametrization set.

30. The largest release of strain in a bicyclic alkane in Figure 2.15 can occur in bicyclo[2.1.0]pentane:

Strain release = 57.3 – 5.9 = 51.4 kcal/mol

Bicyclo[2.2.0]hexane is a close second:

Strain release = 50.7 - 0 = 50.7 kcal/mol

Note that if we include all bicyclic compounds, a new winner emerges:

Strain release = ~68 - 0 = ~68 kcal/mol

If we include tricyclic compounds, tetrahedrane can release the most strain:

Strain release = ~140 - 66.5 = ~73.5 kcal/mol

31. An alternative explanation is that Cy_2P-PCy_2 (and $Cy_2SiHSiHCy_2$) experience geminal repulsion between the cyclohexyl groups, leading to preference of the gauche conformers in much the same way as in tetraalkylethanes.

32. To calculate the fraction of octane molecules that contain at least one gauche bond, we must consider both the enthalpy and entropy of the process. We will assume that any gauche bond increases the enthalpy by 0.9 kcal/mol. Concerning entropy, there are five bonds that can be gauche, and each has two possible gauche conformations. Therefore, the "degeneracy" of the process is 10.

possible gauche

$\Delta G° = \Delta H° - T\Delta S°$

$\quad = \Delta H° - T(R\ln(n))$

$\quad = 0.9 \text{ kcal/mol} - (298 \text{ K})(0.001987 \text{ kcal/mol K})(\ln 10)$

$\quad = -0.46 \text{ kcal/mol}$

$K = e^{-\Delta G°/RT}$

$\quad = e^{-(-0.46 \text{ kcal/mol})/(0.001987 \text{ kcal/mol K})(298 \text{ K})}$

$\quad = 2.19$

% gauche = 100% × (2.19/(2.19 + 1)) = 69%

A modified group increment for $[C-(H)_2(C)_2]$ that represents a pure anti conformation should be more negative than -4.93, reflecting the greater stability of the anti conformation.

33. The BDE for $CH_3CH_2O\text{-}H$ is $\Delta H°_{rxn}$ for the following reaction:

$$CH_3CH_2OH \longrightarrow CH_3CH_2O\bullet + H\bullet$$

$$BDE(CH_3CH_2O\text{-}H) = \Delta H°_{rxn} = \Delta H_f°(CH_3CH_2O\bullet) + \Delta H_f°(H\bullet) - \Delta H_f°(CH_3CH_2OH)$$

We can calculate $\Delta H_f°(CH_3CH_2O\bullet)$ and $\Delta H_f°(CH_3CH_2OH)$ through group increments. We can obtain $\Delta H_f°(H\bullet)$ through consideration of the BDE of H_2:

$$H_2 \longrightarrow 2H\bullet$$

$$BDE(H_2) = \Delta H°_{rxn} = 2\Delta H_f°(H\bullet) - \Delta H_f°(H_2)$$

$$104.2 = 2\Delta H_f°(H\bullet) - 0$$

$$\Delta H_f°(H\bullet) = 52.1$$

$$\Delta H_f°(CH_3CH_2O\bullet) = \Delta H_f°[C–(H)_3(C)] + \Delta H_f°[C–(O\bullet)(C)(H)_2]$$

$$= -10.20 + 6.1 = -4.10$$

$$\Delta H_f°(CH_3CH_2OH) = \Delta H_f°[C–(H)_3(C)] + \Delta H_f°[C–(O)(C)(H)_2] + \Delta H_f°[O–(C)(H)]$$

$$= -10.20 + (-8.1) + (-37.9) = -56.2$$

Now we can plug into our first equation:

$$BDE(CH_3CH_2O\text{-}H) = -4.10 + 52.1 - (-56.2) = 104.2$$

This value, as expected, is nearly identical to $BDE(CH_3O\text{-}H)$, which is 104.4 kcal/mol.

34. The trend is explained by $A^{1,3}$ strain between R and the H on the carbon β to the carbonyl, which clearly gets more important as R gets bigger. The preference for *s-trans* when R = CH_3 can be explained by a similar interaction involving the carbonyl O in the *s-cis* conformer.

35. Since the *A* value is greater for CH_3 (1.74) than for CO_2H (1.4), the carboxyl groups take less energy to place axial. Therefore, the left conformer should be preferred. An important additional effect that favors this conformer is the possibility of hydrogen bonding between the axial carboxyl groups.

36.

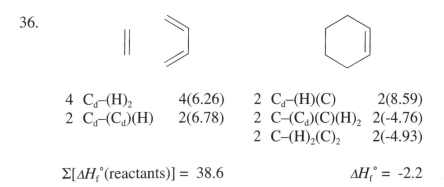

4 C_d–$(H)_2$	4(6.26)	2 C_d–$(H)(C)$	2(8.59)
2 C_d–$(C_d)(H)$	2(6.78)	2 C–$(C_d)(C)(H)_2$	2(-4.76)
		2 C–$(H)_2(C)_2$	2(-4.93)

$\Sigma[\Delta H_f^\circ (\text{reactants})] = 38.6$ $\Delta H_f^\circ = -2.2$

So $\Delta H = \Sigma[\Delta H_f^\circ (\text{products})] - \Sigma[\Delta H_f^\circ (\text{reactants})] = -2.2 - 38.6 = -40.8$ kcal/mol

Using bond dissociation energies, we consider only the bonds that are formed or broken:

$\Delta H = \Sigma[BDE(\text{bonds broken})] - \Sigma[BDE(\text{bonds formed})]$

$= 2BDE(C=C) - 4BDE(C-C)$

$= 2(174) - 4(90) = -12$ kcal/mol

In both cases, the reaction is found to be exothermic, though the values differ by a significant margin.

37. As p character increases in the bonding hybrids, the directionality of the hybrids increases, the tolerance of non-optimum bond angles decreases, and the bending vibrations become stiffer. A reinforcing effect is that as the number of groups bonded to the central C increases, angle variations will lead to greater repulsions between these groups.

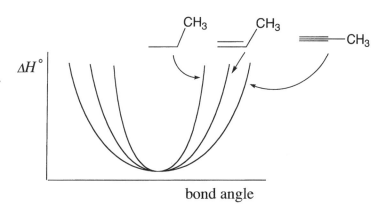

bond angle

38.

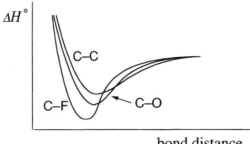

bond distance

As the electronegativity difference between the bonded atoms gets larger, the bond gets stronger and shorter. Also depicted is the steepening of the potential curve as the bond gets stronger. Note that the differences shown are somewhat exaggerated.

39.

$$v = \frac{1}{2\pi}\sqrt{\frac{k}{m_r}} = \frac{1}{2\pi}\sqrt{\frac{4.5 \text{ mdyne/Å}}{\left[\frac{(12.0 \text{ g/mol})^2}{2(12.0 \text{ g/mol})}\right]}\left(\frac{10^{-3} \text{ g cm/s}^2}{1 \text{ mdyne}}\right)\left(\frac{10^8 \text{ Å}}{\text{cm}}\right)\left(\frac{6.022 \times 10^{23}}{\text{mol}}\right)}$$

$$= \frac{1}{2\pi}\sqrt{4.517 \times 10^{28}/\text{s}^2}$$

$$= 3.382 \times 10^{13} \text{ s}^{-1}$$

$$E = hv = (6.626 \times 10^{-34} \text{ J s})(3.382 \times 10^{13} \text{ s}^{-1})\left(\frac{1 \text{ kcal/mol}}{4184 \text{ J}}\right)\left(\frac{6.022 \times 10^{23}}{\text{mol}}\right)$$

$$= 3.23 \text{ kcal/mol}$$

$$= 3.23 \text{ kcal/mol}\left(\frac{349.75 \text{ cm}^{-1}}{\text{kcal/mol}}\right)$$

$$= 1128 \text{ cm}^{-1}$$

This value is within the range given in text for a C-C bond (700-1250 cm^{-1}).

40. A good question to answer first is, "How many different twist-boats are possible in methylcyclohexane?" The twist-boat of cyclohexane has D_2 symmetry, meaning that it has three mutually perpendicular two-fold rotation axes.

Two of the axes are shown for the right-hand view, and the third is the line of sight, perpendicular to the page. (Using a model, we can at least see that the twist boat is close to this ideal geometry – by twisting the flagpole positions of a boat conformation away from each other, we find that two other H's on the other face of the ring are pushed into similar positions.) Since this structure has three different types of H, as labeled in the structure, we should expect three different twist-boats in methylcyclohexane. The Me group can occupy each of the three unique positions.

Of the three possible twist-boats, the one with Me replacing the axial-like H_a is clearly the least favorable, due to flagpole-flagpole van der Waals interactions and *gauche* relationships of Me with both ring C–C bonds. The conformer with Me replacing H_b is next, having one *gauche* and one *anti* relationship with the ring C–C bonds. The remaining conformer, with Me in the equatorial-like position of H_c, has the lowest energy with two *anti* relationships.

Our next question might be, "Can the most stable twist-boat conformer be converted directly to either the axial or equatorial chair form?" Again models are helpful to see the pathways, and the answer is "yes." In the most stable twist-boat, the Me group is already equatorial, so we can get to the equatorial chair by moving the opposite ring carbon first into the plane, giving the half-chair transition state and then further to the chair form. To reach the axial conformer, the Me-bearing carbon should move through the plane.

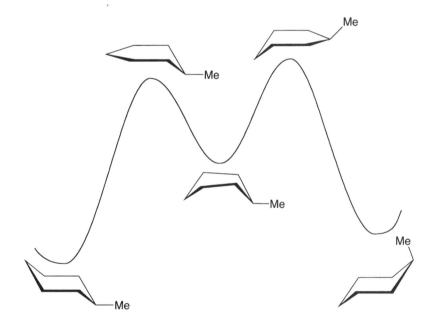

G O I N G   D E E P E R

To determine whether this two-step mechanism from equatorial to axial is really the lowest energy path, we need to consider the relative energies of the possible half-chair transition states. Indeed, given that the axial Me in the chair form comes with an energy cost of only 1.8 kcal/mol, we should expect that the twist-boat forms and the boat transition states between them will all be lower in energy than the half-chair transition states that are 5 kcal/mol or so above the twist-boats (see Figure 2.12). This is a situation in which the Curtin-Hammett principle applies (Section 7.3.3). In brief, this principle says that the relative energies of the twist-boat conformers will have no impact on the relative rates of the processes that require the system to go through the significantly higher half-chair transition states. Though this simplifies the problem in one sense, analysis of the transition states is a somewhat more daunting task, given that there are eight possible half-chairs (not counting enantiomers).

We can cut the number of half-chairs to consider in half by noting that we can avoid Me eclipsing interactions by placing the Me on either the out-of-plane C or the one adjacent to it. In the diagram above, the left-hand transition state is clearly the lowest-energy half-chair, having a fully staggered and equatorial Me group. The right-hand transition state is presumably one of the higher-energy half-chairs, having two H's eclipsed with the Me. The two better candidates for the transition state leading to the axial chair form are the half-chairs with an axial Me on either the out-of-plane C or the one next to it:

Two factors favor the left structure. First, its Me group is fully staggered, while the Me in the right structure is nearly eclipsed with one H. Second, the 1,3-diaxial Me-H interactions of the axial chair form are both relieved in the left structure, while one still persists in the right structure. If we take the left structure as the lowest energy transition state from the axial chair form, we can see that it will lead us to the highest energy twist-boat form – the one with a flagpole Me. With the aid of a model, we can then predict the following as the lowest-energy path from the equatorial to the axial chair form – a path that goes through two half-chairs, two boats, and all three twist-boat forms!

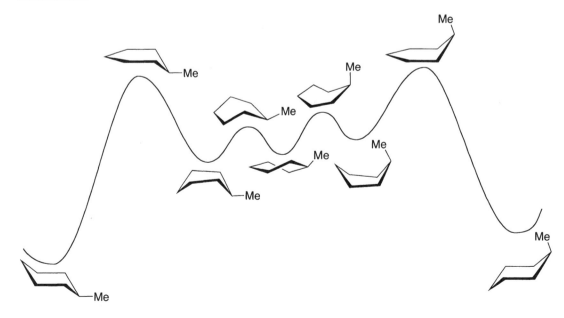

41. $C_4H_8 + 6O_2 \longrightarrow 4CO_2 + 4H_2O$

$\Delta H°_{rxn} = 4\Delta H_f°(CO_2) + 4\Delta H_f°(H_2O) - \Delta H_f°(C_4H_8) - 6\Delta H_f°(O_2)$

$-649.5 = 4(-94.05) + 4(-68.32) - \Delta H_f°(C_4H_8) - 0$

$\Delta H_f°(C_4H_8) = 0.0$

From group increments:

$\Delta H_f°(C_4H_8) = \Delta H_f°[C_d-(H)_2] + \Delta H_f°[C_d-(H)(C)] + \Delta H_f°[C-(C_d)(C)(H)_2] + \Delta H_f°[C-(H)_3(C)]$

$= 6.26 + 8.59 + (-4.76) + (-10.20) = -0.11$

The two values match very closely.

42. From Figure 2.17, the cis isomer of 1,2-di-t-butylethene is destabilized by 9 kcal/mol relative to the trans isomer. In other words, $\Delta H_f°$ is 9 kcal/mol higher for the cis isomer. The same figure shows that the trans isomer is not destabilized relative to *trans*-2-butene – or more correctly, if there is any destabilization in the alkene due to the t-butyl groups, the destabilization is same as that experienced by the alkane (the hydrogenation product) relative to butane. Therefore, the correction for two cis t-butyl groups should be 9 kcal/mol.

Though it is more work, we can also arrive at the correction through comparison of the experimental heat of hydrogenation with the value predicted by group increments. In order to get the right answer, we must remember to include the gauche corrections for the product alkane. There are necessarily four gauche interactions, two for each t-butyl group.

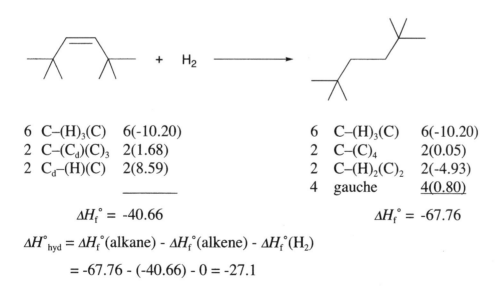

6	$C-(H)_3(C)$	$6(-10.20)$		6	$C-(H)_3(C)$	$6(-10.20)$
2	$C-(C_d)(C)_3$	$2(1.68)$		2	$C-(C)_4$	$2(0.05)$
2	$C_d-(H)(C)$	$2(8.59)$		2	$C-(H)_2(C)_2$	$2(-4.93)$
				4	gauche	$4(0.80)$

$$\Delta H_f° = -40.66 \qquad\qquad \Delta H_f° = -67.76$$

$\Delta H°_{hyd} = \Delta H_f°(alkane) - \Delta H_f°(alkene) - \Delta H_f°(H_2)$

$= -67.76 - (-40.66) - 0 = -27.1$

But the experimental $\Delta H°_{hyd}$ is 36 kcal/mol, so the correction needed in the group increment calculation for the cis t-butyl groups is 9 kcal/mol.

43. The rotation has a 15-fold barrier:

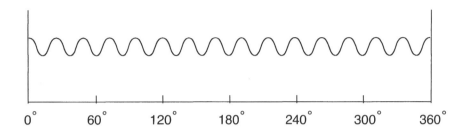

With such a high foldedness, the barrier would be expected to be very small (and it is).

44. At the simplest level, we can deal with the two ends of the main chain independently. In this case, there are only two different staggered conformations:

anti-periplanar   *syn-clinal*

The third staggered conformer is just the enantiomer of the *syn-clinal* conformer shown. While the *anti-periplanar* conformer is expected to be slightly lower in energy, since the Et group is larger than Me, the *syn-clinal* conformation should be much more important than it would be without the geminal Me groups. Thus, relative to an unsubstituted chain, a chain with geminal Me groups is much more likely to be bent, and this should accelerate any cyclization reaction involving functional groups at the ends of the chain. This is the Thorpe Ingold effect.

45. Just as for propene, apparently one completely eclipsed interaction is better than two nearly eclipsed interactions. The oxygen atom in acetaldehyde is sterically smaller than the CH_2 group in propene, so the preference is smaller.

vs.

46. *t*-Butyl radical has two more methyl groups than *n*-butyl radical, and the difference between the CH_3 and CH_2 group increments outweighs the energy difference between the radical group increments. However, one should be careful when comparing group increments that include different atoms. Thus, the statement that the [•C–(C)(H)$_2$] increment, representing a CH_2 group, is less destabilizing than the [•C–(C)$_3$] increment, representing a C atom, is suspect. (See also exercise 23a.)

47. The strain in this alkene leads to some pyramidalization of the alkene C's. This should be accompanied by a rehybridization of these C's from sp^2 towards sp^3. Therefore, the p orbitals involved in π bonding should be rehybridized toward sp^3, and this is apparent in the HOMO.

48. The methyl group in 2-methylallyl cation causes a stabilization of 8 kcal/mol relative to allyl cation. We generally consider the charge to be localized on the end C's of allyl, as suggested by resonance structures, and the fact that 1-methylallyl is stabilized by 20 kcal/mol relative to allyl is consistent with this idea. The stabilization in 2-methylallyl would appear to be due to something other than hyperconjugation, since the C–H bonds are away from the charge, and C–C hyperconjugation is precluded by the lack of overlap of the C–C σ orbital with the π orbitals. Perhaps the general donating ability of methyl is responsible.

G O I N G   D E E P E R

Looking at other examples in Table 2.8 provides further insight. Comparing 1-propyl cation with 2-methyl-1-propyl cation shows that the methyl substituent provides stabilization of only 1 kcal/mol. This suggests that the "general donating ability" of methyl is highly dependent on the nature of the cation and not just its distance from the charge. To understand this situation better, we should admit that it is naïve to completely trust the simple allyl resonance picture and its prediction that *all* of the charge resides on the ends. In reality, we should expect that some portion of the charge resides on the central C, and this might explain how the general donating ability of methyl can be more important in allyl than in 1-propyl. We can rationalize the charge on the central C by including a third, minor resonance structure with biradical character, as shown below. This also brings back the possibility of hyperconjugation in 2-methylallyl cation, which suggests that we might consider this species as a protonated trimethylenemethane (TMM). The biradical TMM is discussed in Section 14.5.6.

49. We have no group increments available for carbocations, so our strategy will be to use group increments to calculate $\Delta H_f°$ values for the neutral hydrocarbons (with C–H replacing C$^+$) and then to use HIA values to relate these to the cations.

5	C_B–(H)	5(3.30)
	C_B–(C)	5.51
	C–(C_B)(C)$_2$(H)	-0.98
	C–(H)$_2$(C)$_2$	-4.93
2	C–(H)$_3$(C)	2(-10.20)

2	C_d–(C_d)(H)	2(6.78)
2	C_d–(H)(C)	2(8.59)
	C–(C_d)(C)(H)$_2$	-4.76
	C–(C_d)(C)$_3$	1.68
2	C–(H)(C)$_3$	2(-1.90)
2	C–(H)$_3$(C)	2(-10.20)
	3-mem ring corr.	27.6

$$\Delta H_f° = -4.30 \qquad\qquad\qquad \Delta H_f° = 31.1$$

By definition, the $\Delta H_f°$ values represent the ΔH for formation of each compound from its elements:

By reversing the first equation and adding it to the second, we get

Now consulting Table 2.8, we see that the closest cations available to the ones of interest are *sec*-butyl cation and phenonium cation (protonated benzene, $C_6H_7{}^+$). If we assume that the HIA values for these cations are the same as those for the cations of interest, we can proceed. By definition of HIA,

By reversing the first equation and adding it to the second, we get

$$\Delta H = -35$$

If we now add the equation above that relates the two neutral hydrocarbons, we get

$$\Delta H = 0.4$$

This suggests that the formation of the phenonium ion is essentially thermoneutral.

Though this result is consistent with the experimental observation of such 1,2-phenyl shifts, we should be cautious concerning the accuracy of this calculation. For example, the assumption that the cyclopropyl ring would have no effect on the HIA of the phenonium cation is probably invalid, given the strong stabilizing effect of cyclopropyl substituents on the cationic carbon as is also apparent in Table 2.8. The effect will almost certainly be weaker in this cation, which is already stabilized by resonance, but it would tend to make the bridging reaction significantly exothermic.

50. While pK_a values for the conjugate acids of the anions shown would likely be difficult to find, we can find values for reasonable models.

The carbanion can be approximated as the ethyl anion, whose conjugate acid, ethane, has a pK_a of approximately 50 (Table 2.10). The cyclopropoxide anion can be approximated as isopropoxide, whose conjugate acid, isopropanol, has a pK_a of 17 (Table 5.5). These estimates suggest that based only on the stability of the type of anion, an equilibrium between these two anions would favor the alkoxide by 33 orders of magnitude. We can then put this in energetic terms:

$$\Delta G = -RT \ln K = -(0.00199 \text{ kcal/mol K})(298K)\ln(10^{33}) = -45 \text{ kcal/mol}$$

The structural differences between the anions – bond types and ring strain – must also be accounted for.

$$\Delta H = \Sigma[BDE(\text{bonds broken})] - \Sigma[BDE(\text{bonds formed})]$$

$$= BDE(C = O) - BDE(C - O) - BDE(C - C)$$

$$= 179 - 92 - 90 = -3 \text{ kcal/mol}$$

Ignoring the difference between free energy and enthalpy, the pK_a and bond strengths together favor the product by a total energy of 48 kcal/mol. The ring strain will counter this trend by 28 kcal/mol, giving an overall product preference of 20 kcal/mol. Thus, the product indeed appears to be favored very strongly.

51.

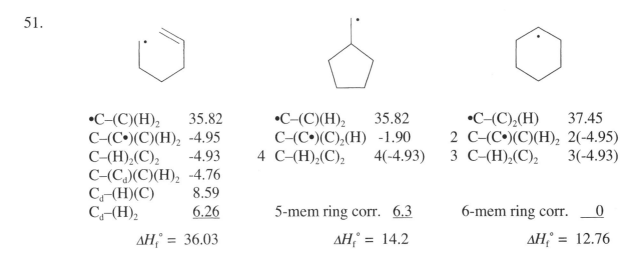

•C–(C)(H)$_2$	35.82	•C–(C)(H)$_2$	35.82	•C–(C)$_2$(H)	37.45
C–(C•)(C)(H)$_2$	-4.95	C–(C•)(C)$_2$(H)	-1.90	2 C–(C•)(C)(H)$_2$	2(-4.95)
C–(H)$_2$(C)$_2$	-4.93	4 C–(H)$_2$(C)$_2$	4(-4.93)	3 C–(H)$_2$(C)$_2$	3(-4.93)
C–(C$_d$)(C)(H)$_2$	-4.76				
C$_d$–(H)(C)	8.59				
C$_d$–(H)$_2$	6.26	5-mem ring corr.	6.3	6-mem ring corr.	0

$$\Delta H_f^\circ = 36.03 \qquad \Delta H_f^\circ = 14.2 \qquad \Delta H_f^\circ = 12.76$$

These results suggest that both ring closures are highly exothermic, but that the observed one is slightly less thermodynamically favored. In fact, the preference for cyclopentylmethyl over cyclohexyl is known to be a kinetic preference.

52. We can see that the radical stabilities are not reflected in the C–Cl BDEs both from the ordering and from the fact that the total range is only 2 kcal/mol. The range in the corresponding C–H BDEs, forming the same series of radicals, is 12 kcal/mol (or 8.5 kcal/mol, based on recently revised values). Since the BDE is the energy required to break the bond,

$$\text{R–Cl} \longrightarrow \text{R•} + \text{Cl•}$$

it depends on the stabilities of all the reactants and products, not just R•. Since Cl• is the same for all four, the only other concerns are effects on the stability of RCl. The small range of C–Cl BDEs suggests that increasing substitution stabilizes RCl as well as R•. Additional methyl groups can stabilize RCl by increasing electron donation toward the electronegative Cl atom. Thus, t-butyl chloride should have a more polar C–Cl bond than methyl chloride. As discussed in Section 2.1.3 (and shown in exercise 21 from Chapter 1), more polar bonds tend to be stronger.

3

Solutions and Non-Covalent Binding Forces

S O L U T I O N S   T O   E X E R C I S E S

1. The preferred geometry is determined by alignment of the dipole of chloroform with the quadrupole of benzene, maximizing the electrostatic attraction. The C-H bond of chloroform should point toward the face of the benzene:

An alternative arrangement places the chlorines in the edge region of benzene. This alignment is not expected to be as good, since the interacting partial charges of both partners are more diffuse.

2. We are asked to calculate a number density, and we can calculate this from the density and molecular weight of water:

$$\frac{1\,g}{mL} \times \frac{1\,mol}{18.02\,g} \times \frac{6.022 \times 10^{23}\,molecules}{mol} \times \frac{1\,mL}{cm^3} \times \left(\frac{1\,cm}{10^8\,\text{Å}}\right)^3 = 0.03342\ molecules/\text{Å}^3$$

The volume of a 20 Å × 20 Å × 20 Å box is 8000 Å³. So, in such a box there are

0.03342 molecules/Å³ × 8000 Å³ = 267 molecules.

3. Reasons that KCl is soluble in water but not benzene:
   i. There is no Born solvation of the K⁺ and Cl⁻ ions in benzene.
   ii. Water is a hydrogen bond donor that can form significant bonds with Cl⁻. Benzene is not a hydrogen bond donor and does not solvate Cl⁻ nearly as well as water does.
   iii. Water molecules are smaller, so that more of them can interact with each dissolved ion. For example, six water molecules can fit around K⁺, while only four benzene molecules can.

4. If X and Y are electron donating substituents, then the carboxylate ion would be further destabilized, leading to lower acidity. Conversely, electron withdrawing substituents would cause the benzene rings to destabilize the carboxylate less (or even stabilize the carboxylate), leading to higher acidity.

5. The dipole moments of ammonia and water are 1.5 and 1.8, respectively (Table 1.5). Therefore, the partial charges are greater in water, leading to a greater electrostatic interaction with benzene.

6. $$E_{sol} = -\left(1 - \frac{1}{\varepsilon}\right)\left(\frac{q^2}{8\pi\varepsilon_\circ a}\right) = -\left(1 - \frac{1}{78}\right)\left(\frac{\left(1.602 \times 10^{-19}\,C\right)^2}{8\pi\left(8.854 \times 10^{-12}\,C^2/J\ m\right)a}\right)$$

$$= -1.139 \times 10^{-28}\ J\ m/a$$

Now we only need to convert units:

$$= \frac{-1.139 \times 10^{-28}\,J\ m}{a} \times \frac{1\,kcal}{4184\,J} \times \frac{6.022 \times 10^{23}}{mol} \times \frac{10^{10}\,\text{Å}}{m}$$

$$= \frac{-164\ kcal/mol}{a/\text{Å}}$$

7. We can just use the equation from exercise 6, plugging in a = 8.5 Å:

$$E_{sol} = \frac{-164\ kcal/mol}{8.5\ \text{Å}/\text{Å}} = -19.3\ kcal/mol$$

8. An aqueous solution of an organic solute is thought to have ice-like regions surrounding the solute molecules. As the solute aggregates, some water is freed from its ice-like structure. Since the heat capacity of water is higher than that of ice, aggregation should tend to increase the heat capacity of the solution. This is opposite from the observation, so there must be other phenomena that override this effect. The explanation is apparently not simple!

9. There are three possible arrangements of three adjacent hydrogen bonds:

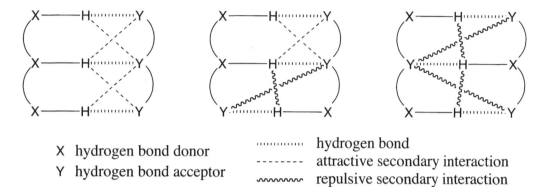

X   hydrogen bond donor
Y   hydrogen bond acceptor

⋯⋯⋯⋯ hydrogen bond
-------- attractive secondary interaction
ⵑⵑⵑⵑⵑ repulsive secondary interaction

Note that the secondary interactions considered are those between the atoms involved in adjacent hydrogen bonds (H and Y). These should be the most important. In the left case, there are four attractive H-Y secondary interactions, and in the right case, there are four repulsive secondary interactions. The G•••C base pair is like the center arrangement, with two attractive and two repulsive secondary interactions.

10. The relevant data is that the *anti:gauche* ratio changes from 70:30 in the gas phase to 55:45 in water. The equilibrium constant in the gas phase is therefore 2.333, allowing calculation of $\Delta G°$:

$$\Delta G° = -RT\ln K = -(1.987 \text{ cal/mol K})(298 \text{ K})\ln(2.333) = -501.6 \text{ cal/mol}$$

In water, the value of K is 1.222, leading to

$$\Delta G° = -(1.987 \text{ cal/mol K})(298 \text{ K})\ln(1.222) = -118.7 \text{ cal/mol}$$

Therefore, the incremental free energy penalty accepted by *anti* butane upon dissolving in water is -118.7 - (-501.6) = 382.9 cal/mol. We can presume that the decrease in surface area upon formation of the *gauche* form stabilizes the surrounding water by an equivalent amount through the hydrophobic effect. The incremental penalty divided by the assumed value for the hydrophobic energy per unit area gives a value for the incremental area:

$$\frac{382.9 \text{ cal/mol}}{40 \text{ cal/mol Å}^2} = 9.6 \text{ Å}^2$$

This calculation suggests that the surface area of *gauche* butane should be approximately 10 Å^2 smaller than that of *anti* butane, or approximately 117 Å^2.

11. This can be most conveniently answered by referring to Eq. 3.17.

$$\Delta G_{rxn} = \mu_B - \mu_A$$

When pure A is first added to the solvent, the driving force, ΔG_{rxn}, is negative, since $\mu_A > 0$ and $\mu_B = 0$. This means that the forward reaction, A $\longrightarrow$ B, is favorable.

12. The expected ranking is:

The first two compounds both have O-H, so the better acid is the better H-bond donor, having the greater positive charge on H. O is more electronegative than N, which is more electronegative than S, explaining the rest of the series.

13. The expected ranking is:

The electronegativity order N<O<F explains the ordering of three of the compounds, as greater electronegativity makes lone pairs less available for hydrogen bonding.

14. The arrangement shown corresponds to $\theta = 0°$, referring to Eq. 3.25. In Chapter 1, the dipole moment and C-Cl bond length of methyl chloride were given as 1.9 D and 1.78 Å. The distance r in Eq. 3.25 is the distance between point dipoles, the position of the points best taken as the "center" of the molecules. With this arrangement of molecules, it does not matter how we define the center. Indeed, the distance between any two corresponding points in the two molecules equals 1.78 Å + 2.3 Å = 4.1 Å, and this is the value we should use for r.

$$E = -\frac{\mu_1\mu_2\left(3\cos^2\theta - 1\right)}{4\pi\varepsilon_o r^3} = -\frac{\mu_1\mu_2\left(3\cos^2(0) - 1\right)}{4\pi\varepsilon_o r^3} = -\frac{\mu_1\mu_2}{2\pi\varepsilon_o r^3}$$

$$= -\frac{(1.9D)(1.9D)}{2\pi(8.85\times10^{-12}C^2/J\ m)(4.1\ \text{Å})^3} \times \left(\frac{3.34\times10^{-30}C\ m}{D}\right)^2 \times \left(\frac{10^{10}\ \text{Å}}{m}\right)^3 \times \frac{kcal}{4184\ J} \times \frac{6.02\times10^{23}}{mol}$$

$$= 1.5\ kcal/mol$$

The energy, inversely related to r^3, clearly gets larger quickly for smaller r. However, this attraction becomes dominated at too small r by repulsions between the electrons around Cl and C and at even smaller r by repulsions between the Cl and C nuclei. So there are other interactions besides dipole-dipole that one must consider. One should also realize that the point dipole assumption implicit in the equation breaks down when the r value is too small.

15. Table 3.5 shows that the diffusion coefficients for Na^+ and Cl^- (1.33 and 2.0×10^{-9} m^2/s) are both larger than that for sucrose (0.52×10^{-9} m^2/s). Since the ions will diffuse through the solution faster, a homogeneous solution will be formed faster from the 1 M NaCl.

16. This criterion represents the equilibrium constant between two geometries. If endothermic steps were discarded, then no barriers would be surmountable in the calculations, no matter what temperature is used for the simulations.

17. The larger α value (Table 3.1) for water relative to methanol (1.17 vs. 0.93) indicates that water is the stronger H-bond donor, while the larger β value for methanol relative to water (0.66 vs. 0.47) indicates that methanol is the stronger H-bond acceptor. We can rationalize this by noting that a methyl group is inductively electron-donating relative to hydrogen. When a water or methanol molecule acts as an H-bond donor, it partially releases a proton, placing a partial negative charge on O. The presence of a donating methyl group would make this less favorable. However, when acting as an H-bond acceptor, the O releases electrons to H, which is more favorable in the presence of a donating methyl group. A statistical effect also favors water as an H-bond donor: it has two H's that it can donate.

18. Acetone has a high dipole moment, so it's heat of vaporization is largely due to dipole-dipole interactions. Benzene has no dipole moment, but it does have a significant quadrupole moment, leading to quadrupole-quadrupole interactions (attractive in the edge-to-face geometry). The high polarizability of benzene also leads to significant induced dipole-induced dipole interactions. While chloroform does have a dipole moment, the moment and the resulting interactions are relatively small.

19. Solvation of the delocalized charge of the low-barrier hydrogen bond (LBHB) will be less effective than solvation of the localized charge that results when the LBHB is cleaved. Also, solvation is somewhat random and dynamic, such that at any particular time, the two ends of the LBHB will be differentially solvated. This will tend to strengthen one H-heteroatom bond at the expense of the other.

20. The aromatic bases in DNA, unlike benzene, have heteroatoms and are involved in H-bonding – both of which produce partial charges. The bases presumably stack in order to arrange these charges to give favorable electrostatic interactions. The rings, like benzene, have quadrupole moments that can participate in the electrostatic interactions. The electrostatics can also be analyzed in terms of dipole-dipole interactions. Another driving force for stacking is van der Waals interactions. Stacking accompanies base pair formation, which is driven by H-bonding and the release of H-bonded water molecules, an entropic effect.

21. The charge-transfer associated with the UV-vis absorption contributes to the binding energy only *after* the absorption and charge-transfer. The electrostatic attraction is then increased between the charged components.

22. One explanation is that solvation, in stabilizing the partial charges on O and N, increases the importance of the zwitterionic amide resonance structure. This increases the C–N π bonding character, which increases the rotation barrier. Another explanation is that the planar amide experiences stronger solvation than the rotation transition state for other reasons. For example, the N methyl groups might sterically interfere with solvation of the π system in the transition state.

An alternative explanation for slower rotation in solution is that associated solvent molecules must be dragged along with the rotating groups. This effect, which might be called "solvent friction," is a dynamic effect in the context of transition state theory in that the argument is based not on the relative energies of the reactant and transition state but instead on the altering of motion over the potential surface.

23. Rotation is not observable for spherical entities.

24. With its more concentrated charge, lithium ion is more effectively solvated than sodium ion. This leads to a larger effective size for lithium, with its associated solvent molecules, and a smaller diffusion coefficient.

25. We might expect that the enol form, being aromatic, should dominate, as it does in phenol. The key in these N-containing compounds is the nature of the N lone pair. In the enol forms, the lone pair lies in the ring plane in an sp^2 orbital and is not in conjugation with the π system. In the keto forms, the lone pair is in a p orbital and part of the ring π system. In fact, this allows the participation of zwitterionic resonance structures that are aromatic. The keto forms, with their higher dipole moments, are stabilized in solution.

4

Molecular Recognition and Supramolecular Chemistry

S O L U T I O N S   T O   E X E R C I S E S

1. Paraquat is a dicationic, electron-poor aromatic compound and the aromatic rings in the polyether cyclophane are very electron-rich. A cation-π interaction should be expected between the cationic guest and the aromatic host. In addition, because the paraquat is very electron-deficient, we should expect π donor-acceptor interactions. The ether oxygen atoms could also interact with the paraquat aromatic rings through polar-π interactions. (All of these interaction types are described in Section 3.2.4.)

2. A large hydrophobic effect explains the large positive $\Delta S°$. Many water molecules involved in hydrogen bonding to both the polyamide and DNA are released upon binding. This is especially true for the DNA double helix, which has a large number of exposed hydrogen bond donors and acceptors that interact strongly with water. Release of these specifically bound water molecules contributes strongly to the $\Delta S°$ value.

3. Each of the three complexes has important secondary interactions in addition to the three primary hydrogen bonding interactions (see Section 3.2.3v and exercise 9 in Chapter 3). The first complex has four repulsive secondary interactions, leading to a lower cooperativity and a lower K_a. In the third complex, the secondary interactions are all attractive, leading to a higher cooperativity and a higher K_a. The second complex has two repulsive and two attractive secondary interaction, explaining its intermediate K_a. Note that this order was predicted in exercise 9 from Chapter 3.

    ⋯⋯⋯⋯   hydrogen bond
    ‑‑‑‑‑‑‑   attractive secondary interaction
    ⌇⌇⌇⌇⌇⌇   repulsive secondary interaction

4. The trend is opposite of what would be expected for just a cation-π interaction. Based on gas phase studies, the smaller RNH_3^+ cation should bind more strongly than the larger $RNMe_3^+$. So the explanation must involve the other choices of the cation guests. In other words, in what other interactions would they participate? Since the solvent is water, RNH_3^+ would be expected to be strongly solvated with hydrogen bonds, while $RNMe_3^+$ would not, and indeed primary ammonium ions show much larger aqueous solvation energies than quaternary ammonium ions. This can explain the host's apparent preference for $RNMe_3^+$ – it is more poorly solvated in water.

5. We can simplify the situation greatly by noting that there are only two specific interactions of interest (the attractive $\Delta G°(NH\text{-}\pi)$ and the repulsive $\Delta G°(H\cdots H)$) and that all others can be lumped together. In fact, $\Delta G_B°$ and $\Delta G_D°$ represent the sums of these other interactions, since B and D do not contain either the attractive or the repulsive interactions of interest. Note that assuming that $\Delta G_B°$ and $\Delta G_D°$ to be equal is an unnecessary approximation. Since B and D have groups that are very sterically different, it is likely that these energies are not equal even if they are similar. We can now define $\Delta G_A°$ and $\Delta G_C°$:

$$\Delta G_A° = \Delta G_B° + \Delta G°(NH\text{-}\pi) + \Delta G°(H\cdots H) \text{ and } \Delta G_C° = \Delta G_D° + \Delta G°(H\cdots H)$$

So $\Delta G°(NH\text{-}\pi) = \Delta G_A° - \Delta G_B° - \Delta G°(H\cdots H)$ and $\Delta G°(H\cdots H) = \Delta G_C° - \Delta G_D°$

Substituting, $\Delta G°(NH\text{-}\pi) = \Delta G_A° - \Delta G_B° - (\Delta G_C° - \Delta G_D°)$

$$\Delta G_A° - \Delta G_B° - \Delta G_C° + \Delta G_D°$$

Note that this analysis does make one approximation: the assumption that $\Delta G°(H\cdots H)$ is equal in complexes A and C. Though the steric effect noted above can have an effect here also, it is hoped that this effect is small.

6. The Benesi-Hildebrand method uses absorbance measurements to determine K_a. The absorbance is potentially due to three different species:

$$A = A_{HG} + A_H + A_G$$

But the experiment is typically done at a wavelength where one of the free components, usually the host, does not absorb. Thus, A_H is zero, and

$$A = A_{HG} + A_G$$

Substituting in Beer's Law ($A = \varepsilon b c$) for both A_{HG} and A_G:

$$A = \varepsilon_{HG} b[HG] + \varepsilon_G b[G]$$

The measurements are each related to the initial measurement (A_0), before any host is added ([H] = [HG] = 0 and [G] = [G]$_0$):

$$\Delta A = A - A_0$$

$$= \varepsilon_{HG} b[HG] + \varepsilon_G b[G] - \varepsilon_G b[G]_0$$

$$= \varepsilon_{HG} b[HG] + \varepsilon_G b([G] - [G]_0)$$

When host is added some of the guest is complexed, so [G]$_0$ = [G] + [HG] and [HG] = [G]$_0$ - [G]. Substituting,

$$\Delta A = \varepsilon_{HG} b[HG] - \varepsilon_G b[HG]$$

$$= b[HG](\varepsilon_{HG} - \varepsilon_G)$$

$$= b[HG]\Delta\varepsilon$$

where $\Delta\varepsilon = \varepsilon_{HG} - \varepsilon_G$. The binding isotherm in Eq. 4.24,

$$[HG] = \frac{[H]_0 K_a[G]}{1 + K_a[G]}$$

was set up under the assumption of constant [H]$_0$. For our purposes, we need the corresponding isotherm where [G]$_0$ is held constant. Since H and G are functionally equivalent, we can just write

$$[HG] = \frac{[G]_0 K_a[H]}{1 + K_a[H]}$$

If we assume that $[H]_0 \gg [G]_0$, we can then take the approximation that $[H] = [H]_0$. Using this and substituting the isotherm into the previous equation gives

$$\Delta A = \frac{b\Delta\varepsilon[G]_0 K_a[H]_0}{1 + K_a[H]_0}$$

Taking the reciprocal of both sides gives

$$\frac{1}{\Delta A} = \frac{1 + K_a[H]_0}{b\Delta\varepsilon[G]_0 K_a[H]_0}$$

$$= \frac{1}{b\Delta\varepsilon[G]_0 K_a[H]_0} + \frac{1}{b\Delta\varepsilon[G]_0}$$

This is the equation used for the double reciprocal plot, from which both $\Delta\varepsilon$ and K_a can be determined.

The amount of excess H needed depends on the accuracy required in the value of K_a. Typically, values are desired with less than about 10% error. To keep the discrepancy between $[H]$ and $[H]_0$ under 10%, we would need to keep $[H]_0$ greater than about $10[G]_0$. This should roughly translate to errors of <10% for K_a.

7. To answer this question, we must remember the binding equilibrium:

$$H + G \rightleftharpoons HG$$

The problem is that we are saturating G, and the equilibrium lies too far to the right. Le Châtelier's Principle tells us that we can shift the equilibrium toward the left by reducing $[H]_0$ or $[G]_0$ or both. If we reduce $[H]_0$ alone, our assumption of $[H] = [H]_0$ will be less valid. So we need to reduce $[G]_0$ (and probably would reduce $[H]_0$ as well).

8. The equation we are to derive, Eq. 4.19, has the general form of Eq. 3.22:

$$\ln(K_a) = \frac{-\Delta H^\circ}{RT} + \frac{\Delta S^\circ}{R}$$

$$R\ln(K_a) = \frac{-\Delta H^\circ}{T} + \Delta S^\circ$$

We can substitute for ΔH° and ΔS° using the expressions from Eqs. 4.15 and 4.18:

$$R\ln(K_a) = \frac{-(\Delta H_\circ + T\Delta C_p{}^\circ)}{T} + (\Delta S_\circ + \Delta C_p{}^\circ \ln T)$$

$$= \frac{-\Delta H_\circ}{T} - \Delta C_p{}^\circ + \Delta S_\circ + \Delta C_p{}^\circ \ln T$$

$$= \frac{-\Delta H_\circ}{T} + \Delta C_p{}^\circ \ln T + (\Delta S_\circ - \Delta C_p{}^\circ)$$

which is Eq. 4.19.

A significant difference in heat capacities between reactants and products is expected when the number or types of vibrational and rotational modes for storing energy are very different. In other words, if the reactants and products have differences either in degrees of freedom or in the force constants associated with the degrees of freedom (normal modes), these should lead to changes in heat capacity. When we say "reactants" and "products" in this context, we must remember that these include solvation spheres.

9. In going from A to B to C, ring constraints are removed, such that C is more conformationally flexible than A. Therefore, C requires freezing out this flexibility in order to achieve the optimum geometry for binding, and it binds more poorly than B or A. In the language of molecular recognition, we should say that A is the most preorganized and C is the least preorganized in the series.

10. The three pyrazyl groups are positioned to allow hydrogen bonds with three of the ammonium hydrogens, as shown. These hydrogen bonds augment the cation-π interaction between the ammonium and the benzene ring. The ethyl groups promote preorganization of the host, since the aromatic substituents will tend to orient in alternating directions to avoid steric contact. This alternation places the pyrazyl groups all *syn*, even in the absence of NH_4^+.

11. We can best answer this question by referring to Figure 4.4, which shows three binding isotherms. The middle isotherm, having $[H]_0 = K_d$, is optimum because we are able to saturate the host, yet we still get most of our data points below saturation. Once the host is saturated, further additions of guest produce little change in [HG], so these data points add little to our determination of binding constant. This is the problem in the upper isotherm, where $[H]_0 > K_d$ and most of the data points are beyond saturation. In the lower isotherm, where $[H]_0 < K_d$, all of the data points are below saturation, but the saturation point is not reached. In effect, a large part of the binding picture is missing, and this leads to a less accurate determination of the binding constant.

12. From Eq. 4.11, it is clear that $\Delta G°_{rxn}$ would be zero if the standard state ($[H]_0$, $[G]_0$, and $[HG]_0$) were set to $1/K_a = K_d$.

13. At a glance, it appears that the binding of cyclohexanediol probably shows positive cooperativity, since the K_a value is approximately 100 times higher than that for cyclohexanol. In order to calculate the cooperativity, we only need compare the $\Delta G°$ of binding with twice the $\Delta G°$ of binding for cyclohexanol. But in fact, two K_a values are given for cyclohexanol, one with a monofunctional host and one with a difunctional host. Which is the more appropriate value to use? We will analyze this question below. For right now, let's just calculate it both ways.

Cyclohexanol

$$\Delta G° = -RT\ln K_a$$
$$= -(0.592 \text{ kcal/mol})\ln(13)$$
$$= -1.52 \text{ kcal/mol (monofunctional)}$$

$$\Delta G° = -(0.592 \text{ kcal/mol})\ln(8)$$
$$= -1.23 \text{ kcal/mol (difunctional)}$$

Cyclohexanediol

$$\Delta G° = -RT\ln K_a$$
$$= -(0.592 \text{ kcal/mol})\ln(1.15 \times 10^3)$$
$$= -4.17 \text{ kcal/mol}$$

So the cooperativity is either 4.17-2(1.52) = 1.1 kcal/mol or 4.17-2(1.23) = 1.7 kcal/mol. The good news is that it doesn't make much difference which K_a value we use – cyclohexanediol shows positive cooperativity either way. (Note that positive cooperativity means stronger binding and therefore a more negative $\Delta G°$!)

G O I N G   D E E P E R

The most appropriate choice to represent the single binding interaction depends on the system being studied. There are both statistical and structural considerations. Let us first consider the binding degeneracies (statistics) in each case. Cyclohexanol can bind in only one way to the monofunctional host but in two equivalent ways to the difunctional host, since there are two equivalent binding sites. Cyclohexanediol can also bind in two equivalent ways to the difunctional host, since the two OH groups can switch places. Therefore, the statistics of binding is matched if we use the difunctional host for both guests, allowing us to calculate cooperativity without worrying about any correction for statistics.

Why, given the two-fold degeneracy, is the binding of cyclohexanol actually *weaker* to the difunctional host? The degeneracy should lead to a lowering of $\Delta G°$ by $RT\ln 2$ or 0.4 kcal/mol. ($\Delta S°$ should be more positive by $R\ln 2$.) Other effects must overwhelm this statistical effect, contributing a positive 0.7 kcal/mol to $\Delta G°$ to give the observed increase of 0.3 kcal/mol. The aromatic ring in the benzyl amide has an extra alkyl substituent in the difunctional host. This donating substituent could increase the electron density at the amide NH, making it a poorer hydrogen bond donor. This should be a very small effect since there is no conjugation between the aromatic ring and the NH – and besides, an alkyl group is a weak donor. A more likely explanation is that the large phenanthroline substituent, at least to some extent, tends to occupy the same space needed by the cyclohexanol. It could be that the cyclohexanol serves to restrict the conformational space available to the large substituent (causing a reduction in the $\Delta S°$ of binding) or it could be that a weak π-π or other interaction in the host is disturbed by the bound cyclohexanol. In either case, the difunctional host would have to pay the same penalty upon binding of cyclohexanediol. Therefore, the structural considerations also suggest that the K_a values from the difunctional host give the cleanest comparison.

While it may seem from this discussion that a difunctional host would be the better choice in any analogous host-guest system, this may not be true in all cases. If the host and/or guests are more complicated or less symmetrical, the considerations described here can be different. One is advised to analyze each system to choose the best model.

14. The answer to this question was noted in the answer to exercise 11. If a significant portion of the binding "picture" is missing, the determination of the binding constant will be less accurate.

15. The fact that Hamilton's barbiturate receptor is a large ring suggests that the desired rotaxane will have a sterically capped rod threaded through this ring. Such a system could be made by several approaches. First, we could start with a complex in which the rotaxane rod will be made from a barbiturate with two "arms" that end in phenoxides. All that would need to be done to finish the rotaxane would be capping the rod with large alkyl groups:

Formation of the complex could be done with the bis(phenol) followed by deprotonation.

An alternative would be to use phenoxide chemistry to close the ring around a pre-formed rod:

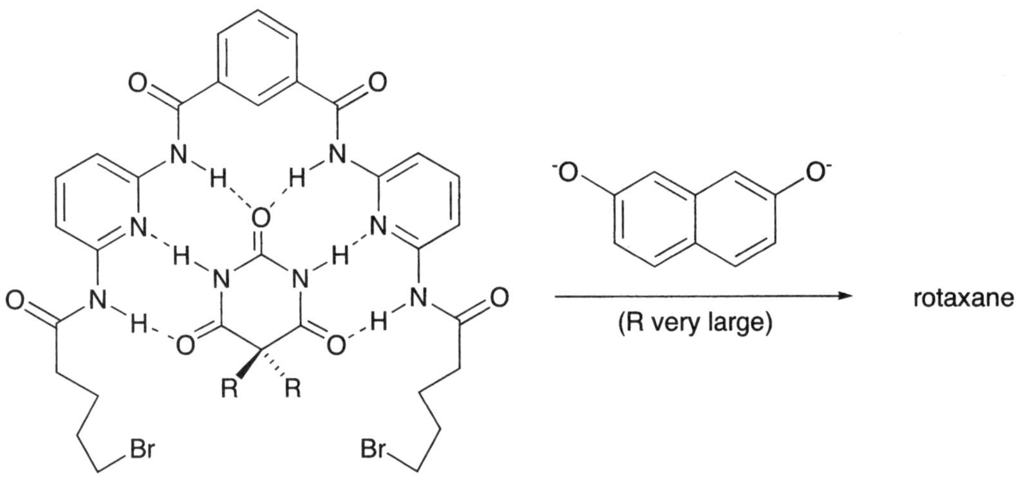

A third alternative would be to bind an appropriate phenoxide into the host cavity, then alkylate the bound phenoxide to form a threaded and capped rod:

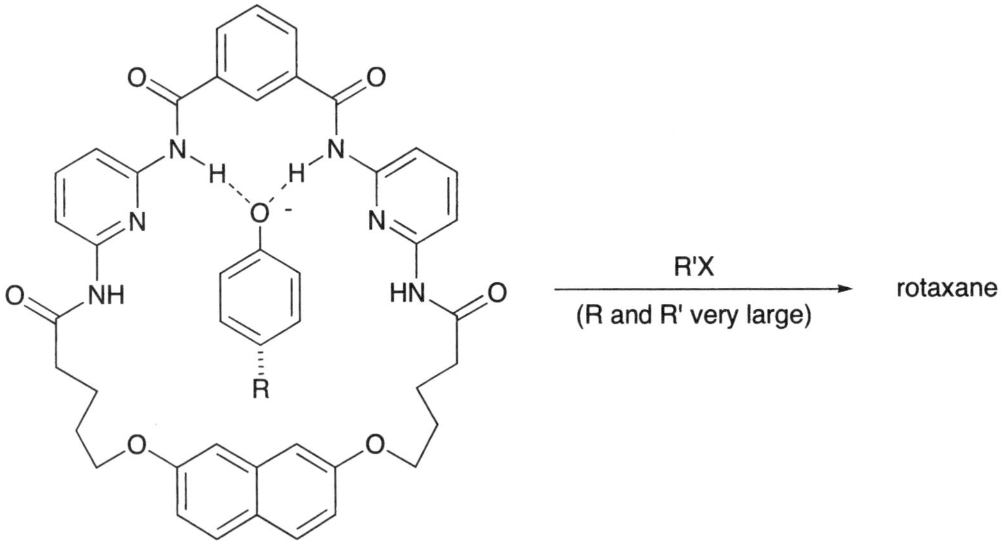

Even though there are fewer hydrogen bonds, the anionic phenoxide is a very good hydrogen bond acceptor, so a complex should form. If the R group is very large, it will have to stick out of the ring in one direction or the other, say away from us. Alkylation will then be blocked from this side and will necessarily occur from the other side. Thus, alkylation of the complex can only give a threaded rod. This rotaxane, unlike the first two, will have a relatively freely sliding ring. This is the strategy used by Vögtle.

16. Reaction rates are usually decreased as temperature is decreased, but this polymerization is different. The reaction involved in the polymerization, a nucleophilic aromatic substitution, would not be expected to be unusual in this respect. However, the fact that the two monomers contain multiple hydrogen bonding groups suggests that an ordered structure is a possibility. In fact, it appears that the polymer strands are self-complementary:

How can this speed up the polymerization? Organization of the polymer after it is formed would not be expected to speed up the polymerization since the organization comes after-the-fact. But if the monomer could be preorganized in a way that reduces the entropy cost of polymerization, a process that forms one molecule from many, this could certainly speed up the polymerization. The larger monomer might be immobilized by the same hydrogen bonding interactions with a polymer strand, and indeed it would be immobilized into a perfect arrangement for polymerization:

At lower temperature, this preorganization would occur more readily (high temperature tends to prevent or destroy order), explaining the increased polymerization rate.

17. The figure implies that a hydrogen bond is important in the binding, and indeed the substituent trend can be completely explained based upon the effect on this hydrogen bond. Electron-withdrawing substituents on the guest phenol will tend to increase the acidity of the phenol and make it a better hydrogen bond donor. Nitro substituents are more withdrawing than cyano substituents, but a nitro at the 3-position is less effective because it is not in conjugation with the phenol oxygen. Electron-donating substituents on the host pyridine ring will tend to increase the basicity of the pyridine, making it a better hydrogen bond acceptor. Dimethylamino substituents are more donating than hydrogen substituents, which are more donating than chloro substituents.

18. If the aromatic guest is bound in the geometry shown for exercise 17, then the new aromatic rings in B and C will be well positioned for favorable edge-to-face interactions with the guest. Though it is difficult to judge from these drawings, one can presume from the higher binding constant for C that the edge-to-face interaction is more favorable in this case. We can rationalize this by noting that the ring in C is probably closer to the guest, being attached at the same end of the naphthalene ring as the hydrogen bond acceptor.

19. A higher K_d value means that a higher concentration of guest is necessary to cause binding, so the lower K_d signifies tighter binding. (Remember that the numerator of K_d contains the concentrations of unbound components, so a higher value means more unbound components relative to the complex.)

20. Given that several minutes are needed to achieve equilibrium, we can be sure that the binding events are slow on the NMR timescale and that separate signals for the boronic acid and ester would be observed (presuming they do not coincidentally overlap). We can get a quantitative sense by using an equation from the Chapter 2 highlight, "The NMR Timescale."

$$k_{coal} = 2.22 \cdot \Delta \nu$$

If the reaction occurs in a time frame of 100 s, the rate constant (k_{obs}) should be on the order of 0.01 s^{-1}. To see what is required for this process to be on the NMR timescale, we can set $k_{coal} = 0.01$ s^{-1}. This gives a value of approximately 0.005 Hz for $\Delta \nu$. So in order for two exchanging nuclei to exhibit coalescence under these conditions, their signals could be separated by no more than 0.005 Hz (roughly 10^{-5} ppm on a 300 MHz spectrometer). First, it is highly unlikely that the signals would happen to appear this close to each other. Even if they did, signals closer than approximately 1 Hz are unresolved in a typical spectrum. So there would be no hope of observing coalescence for such a slow process.

21. To derive Eq. 4.24, we use the relation, $[H]_0 = [H \cdot G] + [H]$, solving for $[H]$ and plugging into Eq. 4.2:

$$K_a = \frac{[H \cdot G]}{[H][G]} = \frac{[H \cdot G]}{([H]_0 - [H \cdot G])[G]}$$

$$[H \cdot G] = K_a([H]_0 - [H \cdot G])[G] = [H]_0 K_a[G] - K_a[H \cdot G][G]$$

$$[H \cdot G] + K_a[H \cdot G][G] = [H]_0 K_a[G]$$

$$[H \cdot G] = \frac{[H]_0 K_a[G]}{1 + K_a[G]}$$

To derive Eq. 4.25, we similarly use the relation, $[G]_0 = [H \cdot G] + [G]$ but solve for $[H \cdot G]$ and plug into the equation just derived:

$$[G]_0 - [G] = \frac{[H]_0 K_a[G]}{1 + K_a[G]}$$

$$[G]_0 - [G] + K_a[G]_0[G] - K_a[G]^2 = [H]_0 K_a[G]$$

$$K_a[G]^2 + (K_a[H]_0 - K_a[G]_0 + 1)[G] - [G]_0 = 0$$

22. The value of K_a is obtained in a similar way for all of the methods. In isothermal calorimetry, the quantity experimentally measured is heat evolved or absorbed, which is directly proportional to ΔH°.

23. To calculate the mole fraction of H•G, we will need Eq. 4.24:

$$[H \cdot G] = \frac{[H]_0 K_a[G]}{1 + K_a[G]}$$

The mole fraction of H•G is just $[H \cdot G]/[H]_0$:

$$X_{H \cdot G} = \frac{[H \cdot G]}{[H]_0} = \frac{K_a[G]}{1 + K_a[G]}$$

As suggested, to get $[G]$ we will use Eq. 4.25:

$$K_a[G]^2 + (K_a[H]_0 - K_a[G]_0 + 1)[G] - [G]_0 = 0$$

We can simplify this equation by taking into account a constraint of this problem:

$$[H]_0 = K_d = \frac{1}{K_a}, \text{ so } K_a[H]_0 = 1 \text{ and}$$

$$K_a[G]^2 + (2 - K_a[G]_0)[G] - [G]_0 = 0$$

We can now solve for $[G]$ by using the quadratic equation. For convenience, let $x = K_a[G]_0$.

$$[G] = \frac{-(2-x) \pm \sqrt{(2-x)^2 - 4K_a(-[G]_0)}}{2K_a}$$

$$[G] = \frac{x - 2 \pm \sqrt{(4 - 4x + x^2) + 4x}}{2K_a}$$

$$[G] = \frac{x - 2 \pm \sqrt{x^2 + 4}}{2K_a}$$

By either inspection or by trying values for x, it is apparent that $x - 2 < \sqrt{x^2 + 4}$, meaning that we should replace $\pm$ with $+$ in order to obtain positive values for $[G]$. Since the equation for $X_{H \cdot G}$ requires $K_a[G]$ in both places that $[G]$ appears, we write

$$K_a[G] = (x - 2 + \sqrt{x^2 + 4})/2$$

The calculation is most easily done by plugging in values of x to get $K_a[G]$ and then computing $X_{H \cdot G}$ by using the equation derived above. Since $[G]_0$ is required to range from $[H]_0/100$ to $100[H]_0$ and $[H]_0 = 1/K_a$, $x = K_a[G]_0$ is required to range from $1/100$ to 100.

x (= $[G]_0/[H]_0$)	$K_a[G]$	$X_{H \cdot G}$
0.01	0.005	0.005
0.1	0.051	0.049
1	0.618	0.382
10	9.099	0.901
100	99.010	0.990

When $x = 1$ ($[G]_0 = [H]_0$), we get a value for $X_{H \cdot G}$ that is close to 0.5. To calculate the value of x that gives exactly 0.5, we need only do the calculation in reverse. The answer comes out to be $x = 1.5$. (A mole fraction of 0.5 would be obtained for $[G]_0 = [H]_0$ if $[H]_0 = 2K_d$, such that at equilibrium $[H \cdot G] = [H] = [G] = K_d$.) Plotting our data gives the following plot:

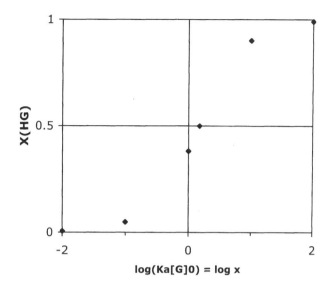

From our table, it appears that each factor of 10 in x below 0.1 corresponds to a factor of 10 in $X_{H \cdot G}$. A similar asymptote occurs for x greater than 10. In other words, the mole fraction of uncomplexed H is reduced by a factor of 10 for each factor of 10 increase in x.

24. In both cases, the iodine atom coordinates to electronegative atoms that have partial negative charges. When attached to a CF_2 group or aromatic ring, the iodine atom should have a partial positive charge. The electrostatic attraction in the complex is enhanced by the high polarizability of iodine, resulting in attractions between the O or N partial charges and an induced dipole on iodine.

25. We can first verify that the concentrations satisfy the equilibrium expression:

$$K_a = \frac{[H \bullet G]}{[H][G]} = \frac{0.84 \text{ mM}}{(9.2 \text{ mM})^2} = 0.010 \text{ mM}^{-1} = 10 \text{ M}^{-1}$$

We should also verify that the concentrations are consistent with the initial concentrations and the reaction stoichiometry.

$$[H]_0 = [H] + [H \bullet G] = 9.2 \text{ mM} + 0.84 \text{ mM} = 10 \text{ mM}$$

$$[G]_0 = [G] + [H \bullet G] = 9.2 \text{ mM} + 0.84 \text{ mM} = 10 \text{ mM}$$

The stoichiometry is also correct. Since $[H]_0 = [G]_0$ and H•G is a 1:1 complex, we should expect that $[H] = [G]$, and both are equal to 9.2 mM.

5

Acid-Base Chemistry

SOLUTIONS TO EXERCISES

1. The reason that the pK_a of H_3O^+ is not zero is that the concentration of water, a constant, is incorporated into the value of K_a (see Eq. 5.7). If instead K_a were taken as K_{eq} in the following equation,

$$K_{eq} = \frac{[H_3O^+][A^-]}{[HA][H_2O]}$$

then substituting $HA = H_3O^+$ and $A^- = H_2O$ would give $K_a = K_{eq} = 1$ and $pK_a = -\log_{10}(1) = 0$. But actually,

$$K_a = \frac{[H_3O^+][A^-]}{[HA]} = [H_2O] = 55.5M$$

$$pK_a = -\log_{10}(55.5) = -1.74$$

2. As more cyano substituents are added, the negative charge in the conjugate base is delocalized over more atoms. The contribution of each cyano group to this delocalization, however, is reduced as their number increases. Therefore, the first cyano group has a larger effect than does the second or third. As mentioned in Section 5.4.4, this effect is called resonance saturation.

3. Using the Henderson-Hasselbalch equation (Eq. 5.12),

$$pK_a = pH - \log_{10}\left(\frac{[A^-]}{[HA]}\right)$$

$$9.2 = 7.2 - \log_{10}\left(\frac{[A^-]}{[HA]}\right)$$

$$\log_{10}\left(\frac{[A^-]}{[HA]}\right) = -2$$

$$\frac{[A^-]}{[HA]} = 10^{-2}$$

This tells us that the conjugate base and acid should be in a 1:100 ratio regardless of the absolute concentration.

4. The conjugate bases of these diketones are the following delocalized anions:

For maximum delocalization, the five-atom π systems should be planar. This would place also the remaining two C's of the five-membered ring and two of the three remaining C's of the six membered ring in the same plane. These arrangements are reasonable in both cases. Considering the acid forms, 1,3-pentanedione is expected to be nearly planar, while this is not expected for 1,3-hexanedione. Therefore, the fact that very little conformational change is required upon deprotonation can explain the higher acidity of pentanedione.

The double Newman projection structures of pentanedione and hexanedione below, representing optimized geometries from molecular mechanics, both support the above reasoning and suggest a refinement. These views show that the carbonyl p orbitals and the acidic C–H bonds are well aligned in pentanedione but are less so in hexanedione. Experimental pK_a values for these compounds are approximately 7 and 10, respectively.

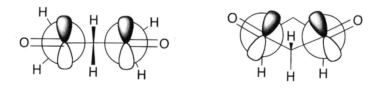

5. Any acid/base equilibrium, such as the general case below, involves two acids (HA and HB) and two bases (A⁻ and B⁻).

$$HA + B^- \rightleftharpoons A^- + HB$$

If HA is the stronger acid, then by definition HA is a better proton donor than HB. In other words, B⁻ is a better proton acceptor (base) than A⁻. So whether viewed as a competition between HA and HB to get rid of the proton or as a competition between B⁻ and A⁻ to take the proton, the result is the same: the equilibrium favors the side with the weaker acid and base. (Note that the weaker acid and base are always on the same side of the equation.)

6. The lower the pH of a solution, the greater its ability to donate protons to a basic solute. If the pH is lower than the pK_a of the conjugate acid of the solute, then the basic solute will be mostly protonated. If the pH and pK_a are equal, the basic solute will be 50% protonated. As the pH continues to rise, the solution's proton donating ability decreases, and the basic solute will not be protonated very significantly if the pH is more than one or two units higher than the pK_a.

7. As discussed in exercise 6, we can determine the protonation state of an acid in solution by comparing its pK_a with the pH of the solution. The pK_as of *p*-nitrophenol and acetic acid are 7.15 and 4.76, respectively (Table 5.5). The pK_a of triethylammonium (the conjugate acid of triethylamine) is not given in Table 5.7A, but the value for trimethylammonium is given: 9.80. The substitution of ethyl for methyl is not expected to have a large effect on pK_a. So in these three cases, we have all three possibilities at a pH of 7.1. *p*-Nitrophenol and its conjugate base, *p*-nitrophenoxide, are present in approximately equal concentration, acetic acid is present predominantly as its conjugate base, acetate, and triethylamine is present predominantly in its protonated form, triethylammonium. If we wished to know the quantitative ratios, we could calculate them easily with the Henderson-Hasselbalch equation (Eq. 5.12).

8. Both Hg^{2+} and S^{2-} are large, soft ions, while O^{2-} is small and hard. The orbitals of soft species tend to be higher in energy than those of hard species, leading to better energy matching in a soft-soft interaction relative to a soft-hard interaction. The sizes of the orbitals are also better matched in a soft-soft interaction, leading to greater overlap. Both of these factors are illustrated in the mixing diagrams below. (The proper choice of atomic hybrid orbitals is debatable, but the factors cited do not depend on this choice.)

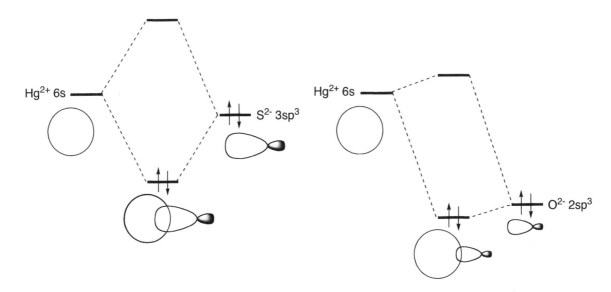

9. A. The second acid should be stronger. The different placement of the methyl groups primarily affects the ability of the ester substituent to be coplanar with the aromatic ring. Coplanarity is required for effective delocalization of the phenoxide charge to the ester, as shown in the resonance structure below. The methyl groups adjacent to the ester in the first compound interfere sterically with the ester, raising the energy of the coplanar geometry. This is called

steric inhibition of resonance. (The methyl placement will also cause a difference in inductive donation from the methyl groups, but this effect should be small in comparison.)

B. The second acid should be stronger. The two acids are tautomers, and the conjugate base of both is the same: phenoxide. Therefore, the acidity only depends on the relative thermodynamic stability of the acids, with the less stable displaying the higher acidity. While most keto-enol equilibria favor the keto form, this one is a strong exception due to the aromaticity of phenol. The actual difference in pK_a values is striking: 10 for phenol and -3 for 2,4-cyclohexadien-1-one.

C. The first acid should be stronger. Like part a, the important factor here is a steric inhibition of resonance in the conjugate base. The methyl groups on N in the second compound prevent (or increase the energy cost for) rotation toward coplanarity of the amine and aromatic ring. This inhibits the resonance delocalization of the N lone pair, such that the amine without the methyl groups is more thermodynamically stable. (Note that the charges on the ammonium cations cannot be delocalized by resonance.) With the nitro groups, these protonated anilines are very strong acids; reported pK_a values are –8.1 and –5.5, respectively.

D. The second acid should be stronger. In the first acid, imidazolium, the positive charge is more effectively delocalized, as the two N atoms take equal shares (see resonance structures). In the second acid, oxazolium, the oxygen takes a smaller share of the charge since it is more electronegative and does not have an attached H to share the charge. Both acids and conjugate bases are aromatic. A statistical effect would tend to make imidazolium more acidic, but this effect is smaller. Imidazolium has two equivalent protons, either of which can be donated to a base. The actual pK_a values for imidazolium and oxazolium are 7.0 and 0.8, respectively.

E. The second acid should be stronger. Several effects can be cited, but the most important is hybridization. An acid with sp² N should be much stronger than one with sp³ N. The same statistical effect mentioned in part D also applies. An important opposing effect is the stabilization of imidazolium by resonance charge delocalization. The actual pK_a values are 11.3 and 7.0, respectively.

10. Hybridization: All alcohols, ROH, have the same O hybridization predicted by VSEPR (though the actual hybridization may vary somewhat). It is also difficult to compare alcohols with differently hybridized atoms attached to O, since, for example, enols are generally less stable than their keto tautomers. Also, strong resonance effects will compete in this case, as for phenol (see below). However, we can readily find examples with hybridization differences two atoms away from O. The acidity ranking of sp > sp² > sp³ can be rationalized as an inductive effect resulting from electronegativity differences (see Section 1.1.8).

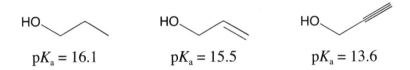

HO⁀⁀⁀   HO⁀⁀＝   HO⁀⁀≡

pK_a = 16.1   pK_a = 15.5   pK_a = 13.6

Electrostatics: The presence of a nearby charge can stabilize or destabilize the negative charge on the alkoxide (the conjugate base). For example, a nearby positive charge increases the acidity:

HO⁀⁀NMe₃⁺   HO⁀⁀

pK_a = 9.3   pK_a = 15.9

Induction: Nearby electron withdrawing groups can stabilize the alkoxide's negative charge, increasing alcohol acidity:

HO⁀⁀CF₃   HO⁀⁀

pK_a = 12.5   pK_a = 15.9

Resonance: Adjacent unsaturation has such strong effects in stabilizing the alkoxide's charge that we consider such species as separate classes of compounds rather than as alcohols.

carboxylic acids   phenols
pK_a ≈ 5   pK_a ≈ 10

Solvation: Steric bulk near a functional group can reduce the ability of solvent molecules to interact with the functional group. Though this is true for both alcohols and alkoxides, the effect is stronger for the charged alkoxides. Therefore, steric bulk reduces stabilization of the alkoxide through solvation, reducing the acidity of the alcohol.

$pK_a = 15.9$ $pK_a = 19.2$

11. In both cases, the pyridine N is more basic because the charge in the conjugate acids can be delocalized onto the other N, as shown, as well as around the ring. The conjugate acids arising from protonation at NH_2 have localized charges due to the already filled N octet and lack of π bonds to N. The difference in acidities in the conjugate acids can be attributed to destabilization of 2-aminopyridinium due to repulsion between the hydrogens on N, each carrying a partial positive charge. In 4-aminopyridinium, the NH and NH_2 groups are far apart, so that this repulsion is not significant. This acid is therefore more stable and less acidic.

12. The two acids A and B have the same conjugate base, so the only difference in the calculation of the two K_as by Eq. 5.7,

$$K_a = \frac{[H_3O^+][A^-]}{[HA]}$$

is whether HA is taken as A or B. In other words, a given solution will have all four species present: A, B, the conjugate base, and H_3O^+, and the four concentrations will provide both K_a values. Since [B] is approximately 10 times smaller than [A], K_a is about 10 times larger for B and pK_a is smaller by about 1. Thus, the pK_a of B is approximately 23.

13. A correlation is easily seen just by comparing the numbers: as the numbers increase in one column, they also increase in the other. A clearer picture of the correlation is obtained by graphing the data, as shown. The correlation is not linear, but it is close. The line shown is from least-squares linear regression. The slope of the line is less than 1 (0.75), indicating that the substituent effects on pK_a are stronger in acetonitrile than in water.

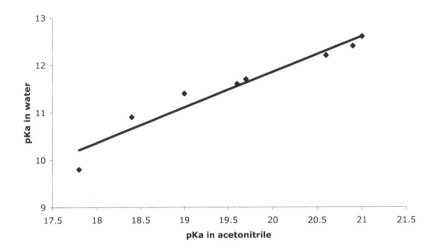

The substituents have a large effect on the pK_as due to both resonance and inductive effects. Resonance structures can be drawn that show delocalization of the positive charge to all three guanidinium N atoms, but delocalization of the charge into the ring is not possible without expanding the octet of the attached N. However, resonance in the benzene ring can transmit charges from the substituents across the ring to the C attached to N, as shown. In this way, a combination of resonance and inductive effects can allow the substituents to influence the pK_a from far away. These effects are stronger in the solvent with the lower dielectric constant (acetonitrile), because solvation of the charges is not as strong.

14. Oxygen, with its high electronegativity, is an inductively electron-withdrawing substituent. In vicinal (1,2) diols, the second oxygen thus stabilizes the negative charge in the alkoxide, increasing the acidity. In geminal (1,1) diols, the second oxygen is even closer, providing even more stabilization. Another effect that also increases acidity is stabilization of the alkoxide through hydrogen bonding:

15. Of the three carboxyl groups in citric acid, the center one should be most acidic due to the inductive effect of the α-OH group. In fact, the pK_a of chloroacetic acid (Table 5.5B) is 2.87, similar to the first pK_a of citric acid. The fact that the third pK_a is higher than normal can be attributed to electrostatic destabilization of the carboxylate due to its close proximity to two other anionic carboxylates.

16. Consider the possible resonance structures for the phosphates:

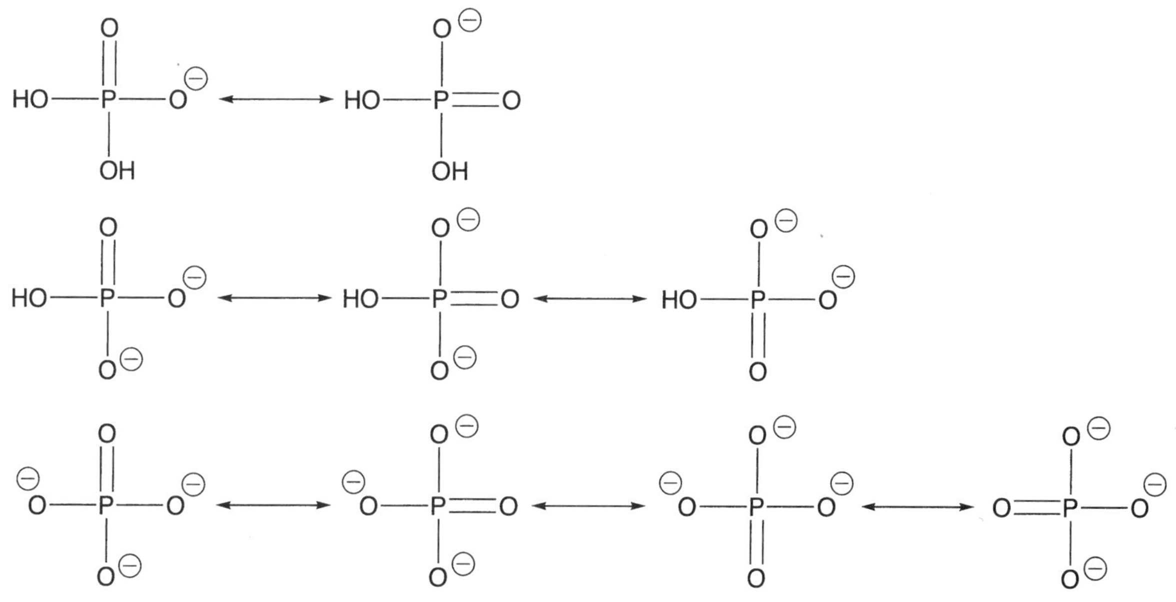

The number of resonance structures, and the number of O atoms involved in the resonance, increases as each proton is lost. One might reason that this would lead to increasing stabilization, such that the increasingly unfavorable electrostatics might be counterbalanced to some extent. The notion of resonance in these anions as sketched above can be disputed on the grounds that the P–O bond is very polar and the π interaction is quite weak, so that P⁺–O⁻ is a better description than P=O. This argument provides a single structure for each ion, as shown below. It is difficult to know what pK_a values one should expect for these two situations, but it does appear from the large pK_a separations that the resonance, if indeed real, does not afford a large amount of stabilization. The high acidity of phosphoric acid can be attributed to the concentration of positive charge on P and its inductive effect on the neighboring OH groups.

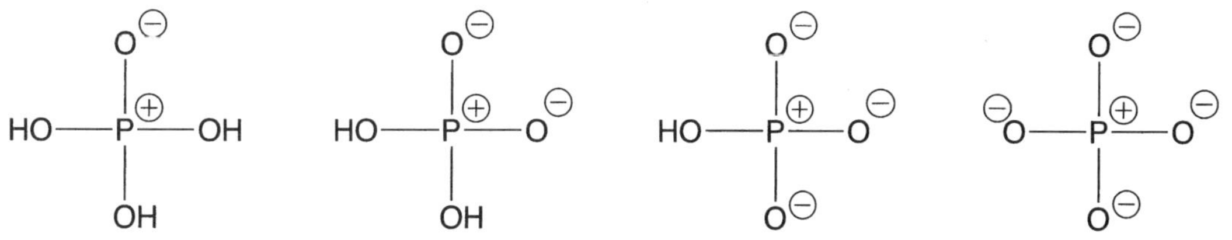

17. The fact that the four acids have similar pK_a values in DMSO is suggestive of a leveling effect. Indeed, Table 5.7B gives the pK_a of the conjugate acid of DMSO as -1.5, not far from the pK_a values for the four acids (0.3 to 1.8). These acids are all acidic enough to protonate the DMSO solvent, such that the measured pK_a values are more reflective of DMSO-H$^+$ than of the solute acids.

18. The *syn* lone pair is more basic. Protonating it gives the more stable *s-trans* conformer of acetic acid, while protonating the *anti* lone pair produces the less stable *s-cis* conformer. In water, hydrogen bonding attenuates the energy difference between the acetic acid conformers. Thus, if the less basic lone pair is protonated, as a consolation, the more basic one is left for hydrogen bonding with water, giving back some of the energy difference. A hydrogen bond involving the acetic acid proton weakens the O–H bond, also attenuating the energy difference between conformers.

19. A. Down (more acidic). The positively charged ammonium ion stabilizes the carboxylate anion, the conjugate base, through both electrostatics and hydrogen bonding.

B. Up. The same stabilizing interactions cited in part a now serve to stabilize the acid (the ammonium cation) rather than the conjugate base (the amine).

C. Up. The face of the benzene ring has a partial negative charge (see Section 1.1.8) that will serve to stabilize the carboxylic acid, through a favorable dipole-quadrupole interaction, and destabilize the carboxylate, through and unfavorable ion-quadrupole interaction.

D. Down. Relative to water, cyclohexane is very poor at stabilizing charges, due to its very low dielectric constant. Therefore, the ammonium (acid) will be destabilized relative to its aqueous solution.

20. From Figure 5.2, we can get H_0 values for the 50% acid solutions. For H_2SO_4, $H_0 = -8.2$. Then, using Eq. 5.19:

$$H_0 = pK_a + \log_{10}\left(\frac{[B]}{[BH^+]}\right)$$

$$-8.2 = -6.2 + \log_{10}\left(\frac{[B]}{[BH^+]}\right)$$

$$\log_{10}\left(\frac{[B]}{[BH^+]}\right) = -2$$

$$\frac{[B]}{[BH^+]} = 10^{-2}$$

$$\% \text{ protonation} = \frac{[BH^+]}{[BH^+]+[B]} \times 100\% = \frac{[BH^+]}{[BH^+]+10^{-2}[BH^+]} \times 100\% = \frac{1}{1.01} \times 100\% = 99\%$$

21. The trend can be explained by considering the conjugate bases:

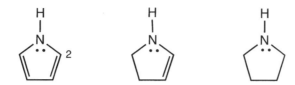

The first base, pyrrole, is an aromatic compound, so the N lone pair electrons are strongly stabilized by resonance. Aromaticity is lost upon protonation. When pyrrole is protonated in strong acid ($pK_a = -4$), protonation actually occurs at C2 rather than at N. Aromaticity is still lost, but the resulting cation is resonance delocalized. The second base is an enamine, affording a much smaller stabilization to the lone pair through resonance. The lone pair in the third base is not stabilized by resonance, and the conjugate acid has $pK_a = 11$.

22. The pK_a of phenol is 10.02 (Table 5.5D) and that of cyclohexanol is likely close to the value for isopropanol: 17.1 (Table 5.5H). The resonance argument is that the negative charge of phenoxide can be delocalized into the benzene ring through resonance. The aryl group is also inductively electron-withdrawing, being attached to O through a more electronegative sp^2 C atom.

CHAPTER

6

Stereochemistry

SOLUTIONS TO EXERCISES

1. Though all 19 chiral amino acids are homochiral (in other words, have the same side chain orientation, as shown below), the priority ordering is different for cysteine. In all other natural amino acids, the side chain has a lower priority than the carboxylate. In cysteine, however, the sulfur atom gives the side chain a higher priority, such that the stereochemical assignment switches to R. (Sulfur has a higher atomic number than oxygen.)

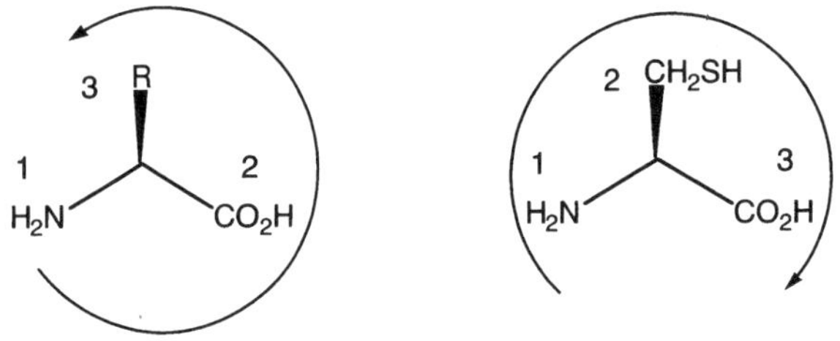

2. A sugar whose standard Fischer projection (with the carbonyl group at or near the top) has a hydroxy group to the right off the bottom stereocenter is designated as D.

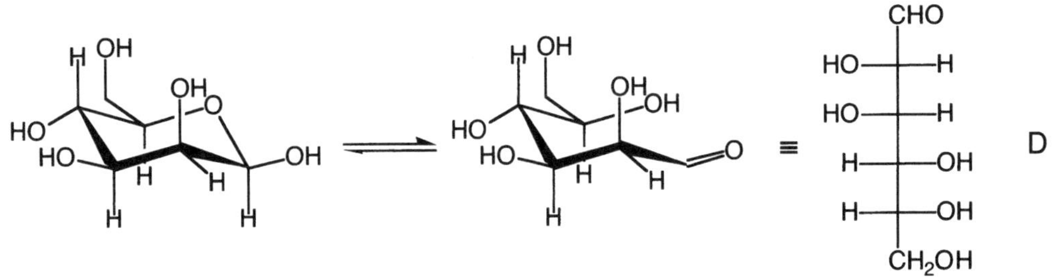

3. To determine Z and E labels, we need to determine the priority order at both ends of the stereogenic alkene.

E. In the top comparison, we find no distinction until the β position: C vs. O.

E. In the top comparison, we must go out three bonds to the γ position to find a difference. The attached atoms are both C, and each of these is attached three times to C. In the next step, we find only H's on the *t*-butyl CH₃ groups while the end C in the 2-propenyl group is attached to two H's and one C. Note that the double bond gives us an "extra" C on each end.

E. In the bottom comparison, Br has a higher atomic number (35) than Se (34).

4. Assigning priorities at both ends of the peptide bond (in bold), we can see that *Z* is the appropriate descriptor. Compared to H, C has higher priority, but compared to O, C has lower priority. So the higher priority groups are on the same side of the bond (*Z*), while the like groups (C) are on opposite sides (trans).

5. A trigonal C with three different ligands is prochiral, since attachment of a fourth different ligand will produce a tetrahedral stereocenter. Attachment from opposite faces will produce opposite configurations of the stereocenter, as shown:

Also important in this case is that attachment of D onto the CH$_2$ group produces a labeled methyl group that is different from the existing methyl.

The *Re* and *Si* faces are assigned below, the labels referring to the faces in view. Priorities of groups attached to a trigonal atom are assigned with the usual priority rules. As described in Section 6.1.2, multiple bonds are treated as multiple ligands, so =CH$_2$ takes priority over methyl (CH$_2$C vs. CH$_3$).

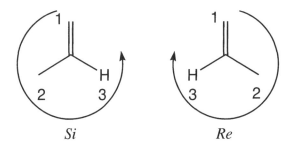

GOING DEEPER

An additional example serves to show that the doubly-bonded ligand does not automatically take a higher priority. In styrene, Ph takes higher priority than $=CH_2$ (CC_3 vs. CH_2C).

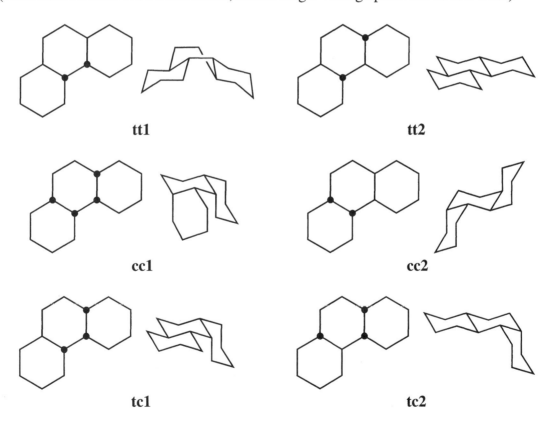

Re   *Si*

6. There are six diastereomers, shown below. One systematic way to classify and find all of these isomers is to note that each of the two ring fusions can be either trans or cis. Therefore, the possible diastereomers are *trans,trans*; *cis,cis*; and *trans,cis*. It turns out that there are two possibilities for each. The dot convention is a simpler way to show the isomers, but it does not convey the three-dimensional shapes very well. When we consider the chair cyclohexane structures, we find that one isomer, **tt1**, cannot have all three rings in the chair conformation. (Note that **cc1** does have three chairs, but no single vantage point shows this well.)

tt1   **tt2**

cc1   **cc2**

tc1   **tc2**

G O I N G   D E E P E R

The answer of six diastereomers might seem surprising if you consider that the maximum number of stereoisomers for a compound with four stereocenters is $2^4 = 16$. Looking at the above six compounds, we can spot two that are *meso* isomers (**tt1** and **cc1**), leaving four that are chiral, representing pairs of enantiomers. (Isomer **cc1** is chiral in the chair conformation shown, but may be considered as a conformationally averaged *meso* isomer, as suggested by the dot-convention structure.) That gives a total of $2 + 4(2) = 10$ stereoisomers. Since each *meso* compound represents two potential "stereoisomers" (*SRSR* and *RSRS* for **tt1** and *RRSS* and *SSRR* for **cc1**), that brings us to a total of 12 of the possible 16 combinations that are accounted for. What happened to the other four?

This apparent conundrum is solved by realizing that there are more possibilities that happen to be identical represented by the two *trans,cis* diastereomers. Just as the end-to-end symmetry makes the *meso* configurations *SRSR* and *RSRS* identical, the chiral structure **tc2** represents the identical possibilities *RSSS* and *SSSR*, along with their enantiomers, *SRRR* and *RRRS*. In this way, each of the two *trans,cis* diastereomers represent 4, not just 2, of the 16 possible configurations. Our accounting is thus complete! (If you still find this confusing, try drawing all 16 possibilities. You will find that 6 of these can be generated from 6 others just by flipping the molecule over, leaving only 10 unique stereoisomers.)

7. For some structures, such as A and B, making the enantiomer is simply a matter of reversing the configuration of *every* stereocenter. For structures C-F, the better approach is to reflect the molecule through an imaginary mirror plane. This naturally reverses any sense of twist and inverts each stereocenter, creating the enantiomer. Note that simply changing the configuration of the stereocenters by switching two groups in E and F would give the enantiomers, but in different conformations. Your drawings need not look exactly like those below to be correct.

8. In A, the biphenyl ring system is viewed as a stereogenic unit as discussed in text; reversing its sense of twist produces a new stereoisomer. The C2 of the 2-butyl substituent is a conventional stereogenic center. The molecule thus has two stereogenic units and four stereoisomers. In B and C, the methyl-substituted carbons are conventional stereocenters. The alkene of cyclohexene might be considered a stereogenic unit in some contexts, but the trans isomer would be so high in energy that we can reasonably think of the cis as the only isomer. The alkenes in the cyclooctadiene, on the other hand, should be considered as stereogenic units, since *trans*-cyclooctenes are stable at ambient temperature.

The central atoms in the square pyramidal (D), trigonal bipyramidal (E), and octahedral (F) structures with all different substituents are all stereocenters. (In fact, switching *any* two ligands produces a stereocenter, though the requirement is only that one such transposition be possible.) In the bicyclo[3.2.1]octane (G), the bridgehead C's may also be considered stereocenters, although again the diastereomers arrived at by switching two ligands would be unreasonably strained. (Such isomers are known for larger bicyclics, however, and are known as *in,out* and *in,in* isomers, having bridgehead hydrogens that point toward the interior of the molecule. The normal isomers, the only reasonable possibilities for smaller bicyclics, are *out,out* isomers.)

9. To determine topicity of the methyl groups, we need to determine if and how they are related by symmetry. This can be difficult to see from drawings on paper. Models can be very helpful, as can redrawing the structure in an orientation that makes the symmetry more obvious. (Models can make this redrawing easier also.) The left structures below are copied from the exercise, and the right structures represent the same molecules viewed from the left, such that the C's in the two five-membered rings appear as perpendicular lines.

Homotopic methyls.

The methyl groups are related by a C_2 rotation axis.

Diastereotopic methyls.

The methyl groups have the same connectivity but are not related by symmetry. We can tell by looking at either structure that they different, since one CH_3 points roughly towards the other, while the other CH_3 points roughly away from the other.

Enantiotopic methyls.

The methyl groups are related by a horizontal mirror plane.

Enantiotopic methyls.

The methyl groups are related by a mirror plane that is roughly the plane of the paper in the left structure and vertical in the right structure.

10. The compound is chiral but not asymmetric. It has a C_2 axis, as shown by the middle perspective, viewed along the C_2 axis. The right drawing is similar to those in exercise 9, drawn from a left-hand viewpoint of the left structure, such that the C's in the three-membered rings appear as perpendicular lines.

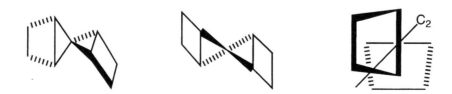

11. Achirotopic points are those that reside on molecular mirror planes or at the center point of a molecule that has an S_n axis. (The vast majority reside on mirror planes.)

All atoms are achirotopic. Viewing down the axis of the C–C bond that links the two rings, we can clearly see two mirror planes that contain all of the C atoms and the H atoms on the rings. (Note that the mirror planes require a 90° dihedral angle between the rings.) The H atoms on the CH_3 groups can also be considered achirotopic, since the time scale for their rotation in and out of the mirror planes is very fast.

All atoms are chirotopic. The presence of a single stereogenic center makes this a chiral compound, such that all atoms are in a chiral environment. There are no achirotopic points in a chiral molecule.

This *meso* compound has a single mirror plane (vertical), so there is a plane of achirotopic points. However, none of the atoms reside on the plane, so all atoms are chirotopic.

This is the *d* or *l* isomer of the last compound. Since it is chiral, no points are achirotopic.

This is another *meso* compound, possessing one mirror plane (horizontal). Like the earlier *meso* example, the mirror plane contains no atoms, all of which are therefore chirotopic.

All atoms are achirotopic. This compound has two mirror planes, each angled at 45° from vertical. These planes contain all atoms except for some of the CH_3 H atoms, which rotate into the planes on a very short time scale.

All atoms are achirotopic, just like the last example, due to two mirror planes angled at 45° from vertical. This molecule has two additional mirror planes (vertical and horizontal) containing additional achirotopic points.

12.

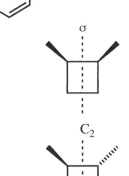

This biphenyl has three axes
and two mirror planes.

No symmetry elements.

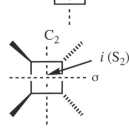

Only one mirror plane.

Only one C_2 axis.

This molecule has one C_2 axis, one mirror plane, and a center of inversion (i). An S_2 axis is equivalent to i, and an axis in any direction will work, as long as it contains the center point.

This molecule has three C_2 axes and two mirror planes. Two C_2 axes are shown; the viewing axis is also a C_2 axis.

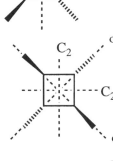

This molecule has one C_4 axis (the viewing axis) and four mirror planes.

13. While it is easy to think of molecules with a C_3 axis and three mirror planes (CH_3F, for example), molecules with a C_3 axis and only one mirror plane are less common. In this case, the mirror plane must be perpendicular to the axis; if parallel, the axis would generate two more mirror planes. Four examples are shown:

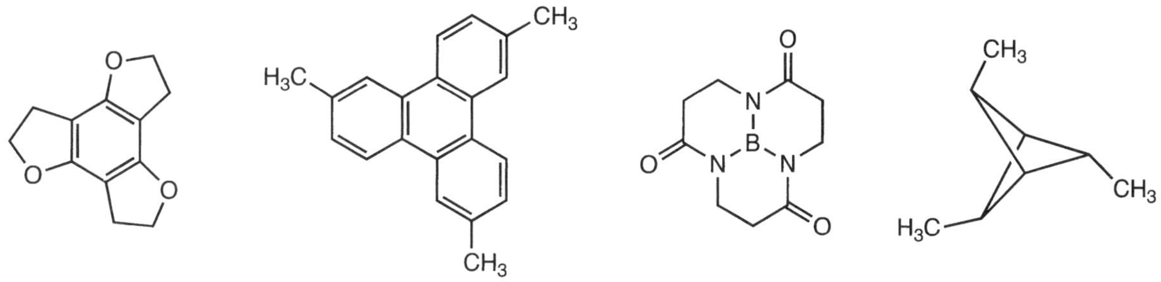

These molecules have C_{3h} point group symmetry, while CH_3F has C_{3v} point group symmetry. Though not included in the scope of this text, point groups are very useful for classification of molecular symmetry.

14. This cyclophane thiol has no symmetry in the conformation shown and is therefore chiral. It can be converted to its enantiomer by rotation about single bonds, but given the observation that its solutions are optically active, we can conclude that the compound is an atropisomer.

Two molecules of this compound can react through two S_N2 steps to give a polycyclic cyclophane product.

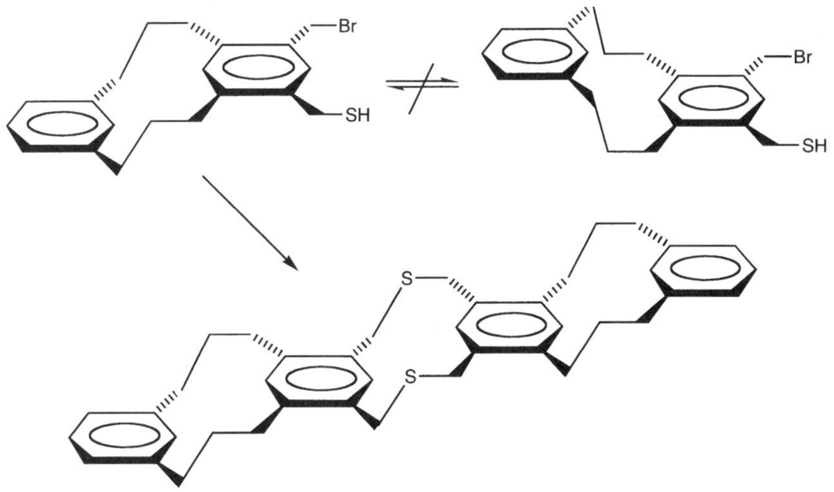

This product, shown both above and below, is achiral. In the lower view, a horizontal mirror plane is apparent. The increase in symmetry, and destruction of chirality, occurs because the two different substituents, CH_2Br and CH_2SH, become equivalent CH_2SCH_2 bridges in the product.

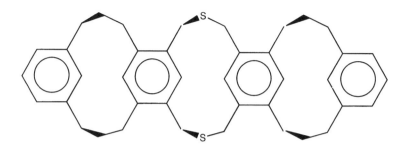

This is a novel process, in that two chiral molecules of the same handedness are combining to make an achiral molecule. The general requirement for a similar result in any system is that the reaction must lead to new symmetry – a mirror plane or other S_n – in the joined product. The reverse process is a molecular realization of a famous parlor trick, "la coupe du roi," in which an apple (achiral) is cut into two identical, chiral pieces. See: F. A. L. Anet, S. S. Miura, J. Siegel, K. Mislow, *J. Am. Chem. Soc.* **1983**, *105*, 1419-1426.

15. An epimer is defined as a diastereomer that differs from a reference compound in the configuration of only one of two or more stereocenters. Therefore, for example, *trans-* and *cis*-decalin are epimers, having two stereocenters and differing at only one. In the tricyclic compound of this exercise, we must change the configuration of two of the four stereocenters to go from the *trans,trans* isomer to the *cis,cis* isomer, so these diastereomers are not epimers.

G O I N G   D E E P E R

If we assign *R* and *S* labels to the isomers above, we can see from the labels alone that the configurations of two of the four stereocenters have been changed.

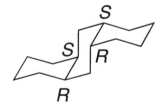

If, however, we go back and consider the simpler case of *trans-* and *cis*-decalin, we have trouble doing the same thing. When we attempt to assign priorities, we find that our rules do not allow us to prioritize the two $(CH_2)_4$ chains that lead to the other stereocenter. So how can the C atoms at the ring fusion be stereocenters if two of the attached groups are the same? In fact, we know that these are stereocenters, because we can interconvert the trans and cis diastereomers by switching two ligands at one of these centers. The answer is that the two $(CH_2)_4$ groups really are different, even though their connectivity is the same. In *trans*-decalin, the two chains are enantiotopic (mirror images, *i.e.*, not the same), and in *cis*-decalin, they are also enantiotopic in several intermediate conformations between the two chair-chair forms (such as the boat-boat conformation shown) as well as in the time-averaged conformation. But with prioritization rules based on connectivity, we are unable to assign R and S labels.

Be assured that those who thought up the prioritization rules took cases like this into account. Although textbooks only give two sequencing rules (atomic number, then atomic mass number to distinguish isotopes), there are indeed three additional rules in the Cahn-Ingold-Prelog (CIP) system. Use of the remaining three rules is not at all trivial, and it is not recommended that you look these up unless you really need them or find such things interesting. If you do need them, they are explained in detail in a 17-page article: V. Prelog and G. Helmchen, *Angew. Chem., Int. Ed. Engl.* **1982**, *21*, 567-583.

Knowing that stereochemical descriptors can be assigned for the decalins leads us to another question. We can reason that the two descriptors for *trans*-decalin must be the same, since the stereocenters are homotopic (interchanged by a horizontal C_2 axis in the drawing above). Likewise, the stereocenters are also homotopic in *cis*-decalin in the boat-boat and time-averaged conformations, so the two descriptors must also be same in this case. Further, the descriptors must be different for the cis and trans isomers, since these are stereoisomers that differ in the spatial arrangement at the stereocenters. But here's the problem: the cis and trans isomers differ by spatial arrangement at only *one* of the two stereocenters (*i.e.*, they are epimers), yet these arguments show that *both* of the descriptors must change in going from trans to cis. How is this contradiction resolvable?

It turns out that both stereocenters in *trans*-decalin are designated r_n, while the stereocenters in *cis*-decalin are s_n. (The lower case descriptors indicate that these are special cases that require rule number 5!) The contradiction is resolved by noting that as a single stereocenter is inverted in going from trans to cis, the descriptor changes not only for that stereocenter, but also for the other. The reason the other one changes has to do with priority rules that are based on things other than connectivity. The moral of this story is that stereochemical analysis of some simple molecules can be complicated. Fortunately, the simple rules work for the vast majority of molecules!

16. Prochiral hydrogens are labeled as "H", and Pro-*S* hydrogens are circled.

17. A 50/50 ratio will be expected if and only if one or more of the species present (including reactants, catalysts, solvent, etc.) are chiral and non-racemic.

Since all species present are achiral, a 50/50 mixture is expected. (The squiggly bond in the product indicates a mixture of configurations.) Another way to think about it: the transition states leading to the two configurations of the new stereocenter are enantiomeric, having the same energy, so the rates of formation of the enantiomeric products are the same. Yet another approach: the faces of the alkene are enantiotopic (related by a mirror plane of the molecule), so there is no preference in the attack of an achiral nucleophile.

Due to the Ph group, the reactant is chiral and the faces of the alkene are diastereotopic. (The mirror plane in the first reactant is destroyed by the Ph.) Thus the attack of any nucleophile on either face leads to diastereotopic transition states (having different energy). The product ratio will not be 50/50.

(Note that we are assuming that the drawing of a single enantiomer actually represents a single enantiomer. In some contexts, the drawing of a single enantiomer can represent a racemic mixture, such as when a chiral product is obtained in an achiral environment, as in the first reaction above. In cases like this, chemists are expected to realize that the compound must be racemic even if a single enantiomer is drawn. In the absence of any notation or context implying that the compound is racemic, the best interpretation is generally that a single enantiomer is implied.)

This is the same chiral reactant and a different nucleophile. By the reasoning above, the product ratio will not be 50/50.

The unsaturated ester is achiral in this case, but the nucleophile is chiral. Even though the alkene faces are enantiotopic, the chiral nucleophile leads to diastereomeric transition states. The product mixture will not be 50/50.

18. This catalyst, in a sense, is an atropisomer. In its most stable conformation it has C_2 symmetry, but it can racemize through rotation of the aromatic ligands such that the Ph groups pass each other. Apparently, this racemization is slow in the active catalyst relative to the polymerization propagation step. Thus, like the C_2 catalysts discussed in the chapter, this one produces isotactic polypropylene, but the sense of the chirality of the catalyst occasionally switches, causing a switch also in the polymer stereochemistry from one block to the next.

19. Both compounds are asymmetric, possessing no symmetry elements. Therefore, *no* hydrogen atoms in these molecules are ever equivalent. Nonetheless, it is a good exercise to prove to ourselves that rotation about bonds is ineffective at making the CH_2 hydrogens equivalent. We can do this by drawing sets of limiting conformations.

The first Newman projection above represents the conformation shown in the exercise, with the CH_2 hydrogens of interest on the front C of the projection. In this conformation, the two H's are clearly different, as one is *anti* to Me and the other *anti* to H. Likewise, in the other two staggered conformations, the two H's are *anti* to different groups. In all three eclipsed conformations, the two H's are eclipsing different groups. It is also clear that rotations around other bonds starting from any of these conformations will not help to make the H's equivalent.

For 3-methylcyclohexene, we could also draw possible conformations. (To be complete, we might include chair, boat, twist-boat, etc.) Fortunately, we can easily see that for each CH_2, one H will always be cis to the Me, while the other will always be trans to the Me for any conformation.

20. To determine topicity of the faces, we need to determine any symmetry relationships between the faces.

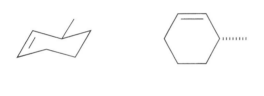

Homotopic faces. Whether a planar or twist geometry is taken, a C_2 axis interchanges the two faces of the alkene. The axis is made obvious by viewing the molecule down the axis. (The faces are also interchanged by a mirror plane in the planar geometry, but this does not alter the homotopic relationship.)

Diastereotopic faces. In any conformation, the Me group is cis to one alkene face and trans to the other. There are no symmetry elements.

Enantiotopic faces. The plane of the paper is a mirror plane that interchanges the faces.

Enantiotopic faces. A horizontal mirror plane in the right view interchanges the two alkene faces.

Enantiotopic faces. The plane of the paper is a mirror plane that interchanges the faces. The *Si* face is in front of the paper, and the *Re* face is behind the paper.

Diastereotopic faces. The only symmetry element is a vertical mirror plane in both views, and this plane does not interconvert the faces.

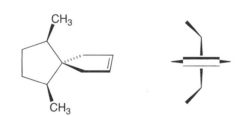

21. The two compounds are redrawn from new perspectives as Newman projections. The left structure shows the view of the original drawing from the top of the page, while the right structure shows the view of the second molecule from the left side of the page. In both cases, the new vantage point makes it clear that the methyl hydrogens are all inequivalent in these static structures. However, two 120° rotations about the C–CH$_3$ bond are sufficient to interchange the hydrogens, placing each of the three hydrogens in the same three locations.

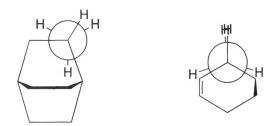

22. A. Both reactions are 100% stereoselective, and the bromination is 100% stereospecific. There is no selectivity for *d* vs. *l* – nor could there be, since both reactants are achiral.

   B. Identical product mixtures are obtained from the isomeric reactants, so there is no stereospecificity. Since the mixtures are 60% *d,l* and 40% *meso*, we can say that there is a 60% stereoselectivity for the *d,l* isomer.

   C. The product mixture for maleic acid is the same as in B: 60% stereoselective for *d,l*. Higher stereoselectivity for *d,l* is observed for fumaric acid: 80%. Since the shift in product percentage with the change in reactant isomer is only 20%, the bromination can be called 20% stereospecific.

   D. Since different diastereomers of the reactant give different diastereomers of the product, each reaction is 100% stereoselective, and the thermolysis is 100% stereospecific.

   E. In this case, the thermolysis is 71% stereoselective but is not stereospecific.

   F. Since the reactant has no stereoisomer (at least not in a practical sense – the *E* isomer is very high in energy, such that it cannot be studied in this reaction), the reaction cannot be stereospecific. The reaction *is* stereoselective. Even though the product isomers are produced in low yield, the stereoselectivity relates only to the selectivity among these products. Therefore, the product percentages should be scaled such that they sum to 100% for computing the selectivity. Thus, the reaction is 65% stereoselective.

23. The possibilities are limitless; three examples are shown. In the first and third molecules, the labeled H's are related by a mirror plane and not by any rotation. In the second molecule, the H's are interchanged by a center of inversion, the only symmetry element of this molecule. In each of the cases shown, there exists one additional pair of enantiotopic H's that are not attached to the same atom.

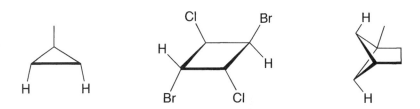

24. The two H's attached to C₃, the back C in the Newman projection shown, are diastereotopic in all conformations. In the conformation shown, for example, one of the H's is anti to OH and the other is anti to H. Indeed, the H's of *any* CH_2 in an asymmetric molecule (including any molecule that has a single tetrahedral stereocenter) are diastereotopic. See exercise 19.

25. It helps to realize first that there are only two stereoisomers for a [2]catenane with directional rings, even if the rings are different (shown as different sizes). These isomers are enantiomeric. We denote the left isomer *R*, since each ring has a clockwise orientation when viewed from the other ring as it enters the center (in the direction indicated by the arrow). Likewise, we denote the right isomer as *S* for its counterclockwise orientation.

a. For a [3]catenane with identical and directional rings, each of the two linkages can be either *R* or *S*. Four permutations of *R* and *S* are possible, just as in molecules with two tetrahedral stereocenters. The first and third structures are enantiomers, but the second and fourth are identical, due to the equivalency of the end rings. This isomer may be considered a *meso* isomer, possessing a topological center of inversion. Therefore, with identical and directional rings, there are three stereoisomers, an enantiomeric pair and a *meso* isomer.

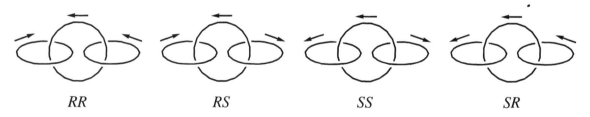

b. For a [3]catenane with different and non-directional rings, three linkage isomers exist, since each of the three rings can be in the center, as shown below. All three topological isomers are achiral, as the plane of the central ring can be a mirror plane.

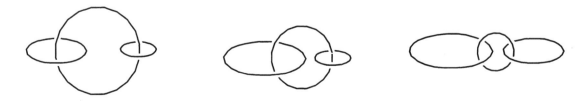

c. For a [3]catenane with different and directional rings, both types of isomerism exist simultaneously. For each of the three linkage isomers of part b, the four directionality isomers of part a would exist, for a total of twelve isomers (six enantiomeric pairs). Note that no *meso* isomer is possible in this case, since the end rings are different.

GOING DEEPER

The exercise instructed us to presume that the individual rings are achiral, and further that the ring plane is a mirror plane. If we remove these restrictions, a new form of topological chirality appears, and the number of possible isomers increases dramatically.

We first allow for the possibility that the individual ring planes are not mirror planes. The rings can still be achiral if a mirror plane perpendicular to the ring plane exists. The presence of such a mirror will also make the ring non-directional, so this change does not apply to parts a or c. An example of such a ring is shown:

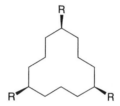

Linking two such rings, whether identical or different, leads to chiral [2]catenanes (at left below), since all mirror planes of the individual rings are destroyed by the linkage. Like the linkage and directional isomerism discussed above, this is a form of topological isomerism, arising from a different type of directionality. For [3]catenanes like those in part b, two diastereomeric pairs of enantiomers will be formed for each linkage isomer. The diastereomers could be called head-to-tail and head-to-head, referring to the relative directions of the end rings.

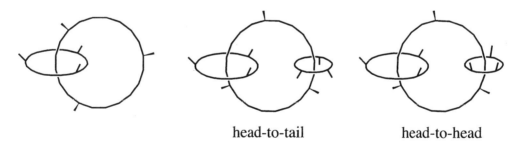

head-to-tail   head-to-head

If the individual rings are chiral, then isomerism arises that is similar to that associated with the presence of multiple stereocenters. If each ring exists as two enantiomers, then a [2]catenane will exhibit two diastereomeric pairs of enantiomers (2^2 isomers), and a [3]catenane will exhibit 4 diastereomeric pairs of enantiomers (2^3 isomers).

Expanding on part c, what happens if we make [3]catenanes with rings that are different, directional in both senses, and chiral? An example of such a ring is shown below. The types of directionality are interdependent, since both senses are switched by flipping the ring over. So the number of possible isomers is $3 \times 4 \times 8 = 96$. Clearly, these issues must be considered when designing a catenane synthesis!

26. It may seem surprising at first that C_{60} has a planar graph, given that there are so many crossed bonds in the diagram. But imagine viewing the molecule from the inside — there would be no crossed bonds! This surrounding shell is not planar, but we can make it planar by stretching bonds and flattening the molecule. Imagine standing inside the sphere and placing your fingers into one of the six-membered rings overhead and pushing to the sides to stretch all six bonds. Keep stretching these bonds while flattening the molecule until this large ring surrounds the rest of the molecule on the floor. A planar graph would be obtained (π bonds omitted):

27. The product would be a [2]catenane.

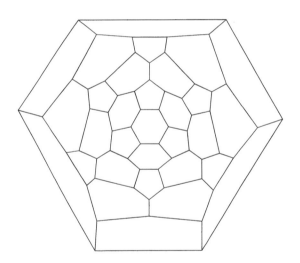

28. In the text, the descriptors were determined by sighting from right to left. Here, sighting from left to right, we obtain the same descriptors:

29. The number of possible arrangements of five different rings in five possible positions is 5! (five factorial). This result from probability theory is easy to understand. There are 5 possible places for the first ring. For each of these, there are 4 remaining possibilities for the second ring, and so on. However, this will lead to a double counting, since, for example, 1-2-3-4-5 is the same as 5-4-3-2-1. So the number of isomers is $5!/2 = 60$.

30. The eclipsed form has a C_5 axis and a perpendicular mirror plane, and in combination these constitute an S_5 axis. The staggered form has a C_5 axis but no perpendicular mirror plane; the C_5 axis is also a S_{10} axis. Both conformations are shown as viewed along the axis.

The Cp–Fe rotation has a five-fold barrier. (One might be tempted to say that ferrocene should have a 25-fold barrier, since it has two five-fold rotors. However, all C's become eclipsed at the same time, reducing the foldedness.)

31. a. We can most easily analyze the symmetry of this molecule by considering the center and the ends of the molecule separately. The center section, shown at the left below, is achiral, having two mirror planes and an S_4 axis. The mirror planes contain the central C–C axis and one of the aromatic rings, and the S_4 axis is coincident with the central C–C axis. The end groups taken together, shown at the right, are also achiral, having a vertical mirror plane. If one of the ends were rotated 180° with respect to the other, there would be a center of inversion (i or S_2).

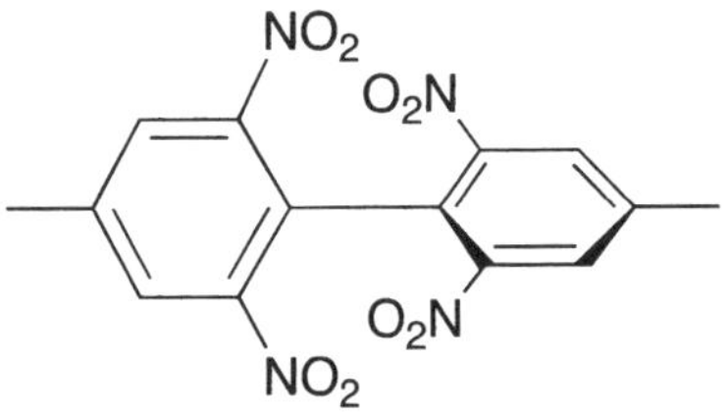

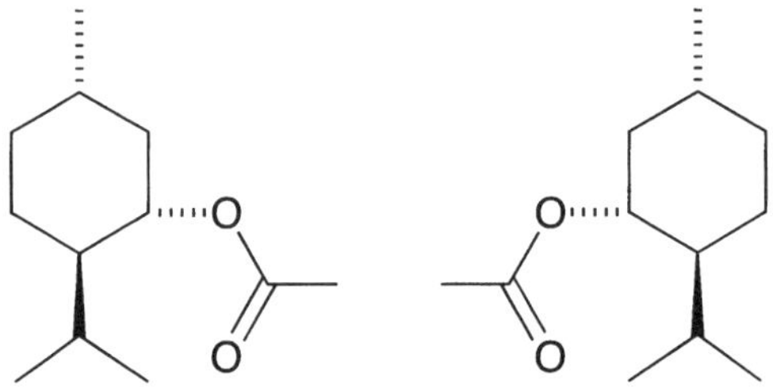

The center and ends of the molecule have no symmetry elements in common. Thus, when the components are put together, there is no symmetry. The molecule is therefore chiral.

b. We can draw the enantiomer by reflection of the molecule through any mirror plane. For convenience, we will choose the plane of the paper (see scheme). We can also convert one enantiomer as drawn to the other by two rotations: a rotation of the entire molecule followed by an internal rotation about two single bonds.

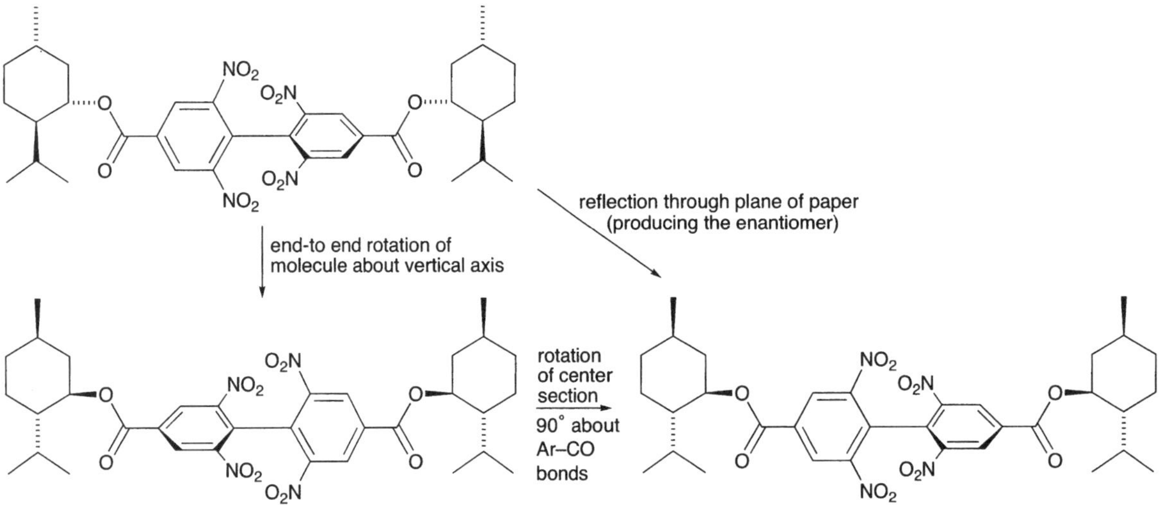

c. Using the reasoning of part a, we can see that the only way this molecule can become achiral, in other words, to possess a mirror plane or other improper rotation, is to rotate about the central C–C bond to make the Ar rings coplanar. This would potentially give the molecule a vertical mirror plane that bisects the central C–C bond or a center of inversion. However, since the nitro groups prevent this from happening, the molecule must always be chiral.

32. For the timescale where ring inversion is slow, we can think of the molecules, for symmetry purposes, as rigidly locked in one chair conformation. For fast inversion, we can treat the molecule as if the six ring carbons lie in a single plane. We should realize, however, that neither the planar conformation nor any other symmetric conformer need participate in the inversion process to enforce the topicities we determine this way (a situation similar to that presented in exercise 31).

Slow inversion   Fast inversion

Diastereotopic (one CH₃ axial, one equatorial but same connectivity)

Enantiotopic (related by mirror plane)

Diastereotopic (one CH₃ axial, one equatorial but same connectivity)

Homotopic (related by C₂ axis – molecule viewed along axis)

Homotopic (related by C₂ axis)

Homotopic (symmetry cannot be decreased by inversion process)

Diastereotopic (one CH₃ axial, one equatorial but same connectivity)

Homotopic (related by horizontal C₂ axis)

33. Replacing an isopropyl CH$_3$ with Br produces a stereocenter. With two such substitutions, the molecule has two stereocenters and $2^2 = 4$ possible stereoisomers: *RR*, *SS*, *RS*, and *SR*. Since the two stereocenters have the same connectivity, an achiral *meso* isomer (*RS = SR*) is a possibility, but in fact, the two stereocenters are not related by symmetry in the geared conformation. (They are obviously different: one points into the back of the other.) Therefore, there are four isomers and all are chiral: two diastereomeric pairs of enantiomers (*RR, SS* and *RS, SR*).

The consequences of correlated 180° rotation of the six groups depends on the isomer. The *RS* and *SR* isomers are interconverted by this process. The *RR* isomer is converted into itself, as is the *SS* isomer. (For the *RR* and *SS* isomers, 180° rotation of the whole molecule about a horizontal axis produces a structure equivalent to that produced by the correlated rotation.)

34. Showing that the metal centers are chirotopic is simple. Since each complex has a C$_2$ axis as the only element of symmetry, each complex is chiral and all atoms in the complexes are chirotopic. To show non-stereogenicity, we must show that stereoisomers are not produced upon switching of any two ligands. The first two complexes, having tetrahedral metal centers, would need to have four different ligands in order to be stereogenic. Both have pairs of identical ligands, and switching any two produces the same complex. The third complex has an octahedral nickel atom with a meridional tridentate ligand. Several possible ligand swaps could produce a diastereomeric complex with a facial tridentate ligand, but this complex would be considerably higher in energy due to the rigidity of the tridentate ligand. So practically speaking, this nickel center is also non-stereogenic. In all three complexes, the stereogenic atoms are carbon atoms in the ligands.

35. This probability can be calculated by realizing first that each product molecule has a 50% chance of being either enantiomer. We will assume a 100% yield – that is, every molecule of starting material is converted to product. (Note that if only one molecule, or any other odd number of molecules, is not converted to product, the probability for an exactly 50:50 ratio will be zero!) This question then becomes a classic problem from probability theory called an

equipartition problem. The probability is the same as the probability that in n flips of a coin, heads will be obtained exactly $n/2$ times.

$$\text{probability of exact } 50:50 \text{ ratio} = \frac{\text{number of } 50:50 \text{ combinations}}{\text{number of total possible combinations}}$$

Let us number the molecules from 1 to n. The numerator above represents all possible ways that half of the molecules could be one enantiomer, say (+). For example, for $n = 2$, there are two possible combinations: either molecule 1 or 2 could be (+). For $n = 4$, there are six combinations: (12, 13, 14, 23, 24, 34). In general, the number of combinations of k items selected from a set of n items is $n!/k!(n-k)!$ In our case, $k = n/2$, so the expression reduces to $n!/((n/2)!)^2$. The denominator above represents all possible combinations, whether 50:50 or not. Since each molecule has two possible states, (+) and (-), that are determined independently of the other molecules, the total number of combinations is 2^n.

$$\text{probability of exact } 50:50 \text{ ratio} = \frac{n!\big/\left(\frac{n}{2}!\right)^2}{2^n}$$

Plugging in some values for n (including 10 and 1000, which were asked for in the exercise):

n	2	4	6	8	10	34	100	340	1000
prob.	0.5	0.375	0.313	0.275	0.246	0.136	0.080	0.043	0.025

Whether using a calculator or computer, we find that we cannot go much further. Both the factorials of the numerator and the exponential of the denominator get very big as n increases, and even with 15-digit number precision (allowing numerator and denominator to approach 10^{308}), the highest n we can use is 1022. However, we can do better by using Stirling's approximation for factorials, which is very accurate for large n:

$$\ln(n!) = \left(n + \tfrac{1}{2}\right)\ln(n) - n + \ln\sqrt{2\pi}$$

To use this approximation, we first take the logarithm of the probability and then substitute, expand, and simplify:

$$\ln(\text{prob. } 50:50) = \ln\left(\frac{n!\big/\left(\frac{n}{2}!\right)^2}{2^n}\right)$$

$$= \ln(n!) - 2\ln\left(\tfrac{n}{2}!\right) - n\ln 2$$

$$= \left(\left(n + \tfrac{1}{2}\right)\ln(n) - n + \ln\sqrt{2\pi}\right) - 2\left(\left(\tfrac{n}{2} + \tfrac{1}{2}\right)\ln\left(\tfrac{n}{2}\right) - \tfrac{n}{2} + \ln\sqrt{2\pi}\right) - n\ln 2$$

$$= n\ln(n) + \tfrac{1}{2}\ln(n) - n + \ln\sqrt{2\pi} - (n+1)\left(\ln(n) - \ln 2\right) + n - 2\ln\sqrt{2\pi} - n\ln 2$$

$$= n\ln(n) + \tfrac{1}{2}\ln(n) - \ln\sqrt{2\pi} - n\ln(n) - \ln(n) + n\ln 2 + \ln 2 - n\ln 2$$

$$= -\tfrac{1}{2}\ln(n) - \ln\sqrt{2\pi} + \ln 2$$

$$= \ln\left(\frac{2}{\sqrt{2\pi n}}\right) = \ln\sqrt{\frac{2}{\pi n}}$$

$$\text{prob. } 50:50 = \sqrt{\frac{2}{\pi n}}$$

This equation reproduces the above-calculated probabilities with increasing accuracy as n increases (2.5% error for $n = 10$ and 0.025% error for $n = 1000$). With this equation, we can directly calculate the probability for $n = 10^{21}$ to be 2.5×10^{-11}. Clearly, the probability of an exact 50:50 ratio will be vanishingly small.

GOING DEEPER

Even though the chance of an exact 50:50 ratio of enantiomers for any laboratory sample is extremely small, the probability that the ratio will be experimentally indistinguishable from 50:50 is very large. Let's first investigate the trend for small n. We will calculate the probability that the ratio will be between 48:52 and 52:48.

A convenient and intuitive way to address this problem is to use Pascal's triangle. This mathematical construct, shown below, is generated line by line by placing "1" on the outside edges and taking each interior number as the sum of the two adjacent numbers above it. It turns out that Pascal's triangle directly shows us the number of combinations needed for our probability calculations. (Completely analogously, Pascal's triangle shows the number of possible spin-state combinations contributing to an NMR multiplet and therefore predicts the relative intensities.)

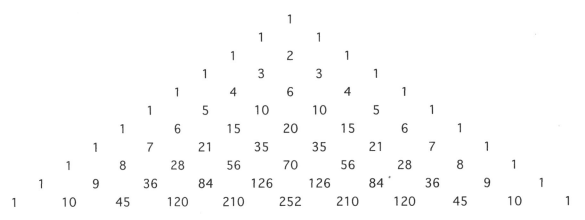

The 1's on the left of the triangle will represent the single combination with all (-) molecules, and the 1's on the right will represent the single combination with all (+) molecules. (There's only one way each can happen, no matter how many molecules there are.) To analyze what can happen with n molecules (or flips of a coin), we look at row $n+1$ of the triangle. For $n = 2$, row 3 shows us that there is 1 way to have zero (+) molecules, 2 ways to have one (+) molecule, and 1 way to have two (+) molecules. The probability of a 50:50 mixture is therefore the center number (2) divided by the sum of the whole row (4), giving a probability of 0.5. For $n = 4$, row 5 shows us that there is 1 way to have zero or four (+) molecules, 4 ways to have one or three (+) molecules, and 6 ways to have two (+) molecules. The 50:50 probability is 6/(1+4+6+4+1) = 0.375. Look at the triangle and note how quickly the numbers near the center get large as n increases. Nonetheless, the sums of the rows get large a bit faster, leading to a decrease in the 50:50 probability as n increases, as seen above.

For neither $n = 2$ nor 4 are there any possibilities to be within the 50±2% range without being exactly 50:50. The first opportunity for this occurs at $n = 50$, where we can have 24 and 26 molecules of each enantiomer. In row 51 of Pascal's triangle, the center number is 1.264×10^{14} and the sum of the row (*i.e.*, 2^{50}) is 1.126×10^{15}, giving a 50:50 probability of 0.112. The numbers on either side of the center, representing the 24:26 and 26:24 combinations, are both 1.216×10^{14}. So the probability of having 50±2% of the (+) enantiomer is $(1.264 + 2(1.216)) \times 10^{14}/1.126 \times 10^{15} = 0.328$, approximately three times that for the 50:50 ratio.

As n gets larger and more possible combinations fall within the ±2% range, the probability increases further. For $n = 100$, the number of (+) molecules can be 48, 49, 50, 51, or 52, and the probability of this is 0.383. For $n = 1000$, the number of (+) molecules can range from 480 to 520, and the probability increases to 0.805. The probability for $n = 10,000$ is 0.99994, while the probability for $n = 100,000$ is 0.999999999999999999999999999999999999. In other words, the probability that the ratio would fall *outside* the 50±2% range would be 10^{-36}, less than your chances at winning the lotto jackpot five weeks in a row (after buying only one ticket each week)!

The answers to this problem may be calculated also by using Stirling's approximation or by using binomial distributions, which for large n are accurately approximated by the normal distribution (see any probability and statistics textbook). Using this method, we can calculate that for $n = 10^{21}$, the probability that the number of (+) molecules will fall outside the 50±0.00000002% range is similar to the five-jackpot probability mentioned above. Thus, we can feel quite assured that in any racemic laboratory sample, the enantiomer ratio will fall close enough to 50:50 that we will be unable to tell otherwise by even our most sensitive of methods.

36. The helicene is *P* (right-handed). The binaphthyl is *M* (left-handed). Note that for a true helix like the helicene, prioritization of groups is not required – the arrow should just follow the helix.

37. Four propeller-shaped stereoisomers exist: two diastereomeric pairs of enantiomers:

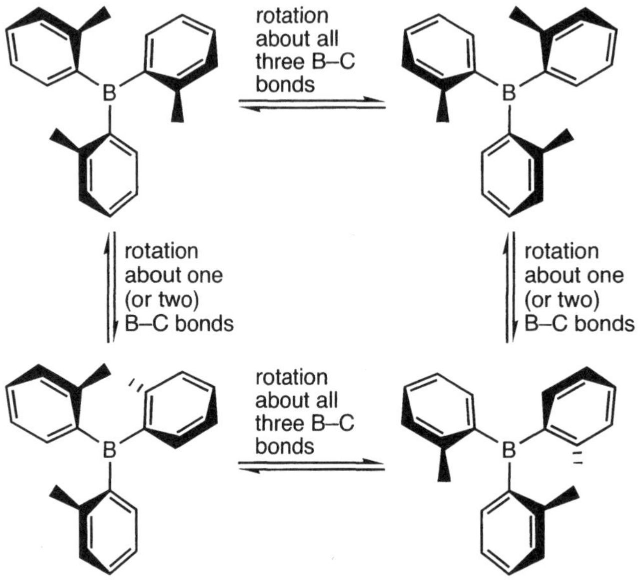

The enantiomers at the top have a C_3 rotation axis, while the two at the bottom have no symmetry. Interconversion of enantiomers requires rotation of all three aryls groups, switching the helicity from *M* (at left) to *P* (at right). None of the rings need rotate through the BC_3 plane, so the racemization is expected to be fast. Interconversion of diastereomers requires rotation of either one or two aryl groups through the plane, and this might be slower due to steric hindrance. For either interconversion, steric crowding can be best avoided through coupled rotations. For racemization, the three rings best avoid each other by rotating at the same time. For interconversion of diastereomers, two rings can best allow the third to rotate through the plane by rotating to become perpendicular to the plane.

38. Since there are no stereoisomers of phenylacetylene, the reaction cannot be stereospecific.

39. We can make use of the synthetic strategy illustrated in Figure 4.15 to make catenanes by coordination of phenanthroline ligands to preorganize the units before cyclization. To make the highly interwoven Borromean rings we have a big organizational job to do, but in principle, each crossing can be accomplished by a separate coordination. Drawing the very large molecules required will be much easier if we use a schematic notation:

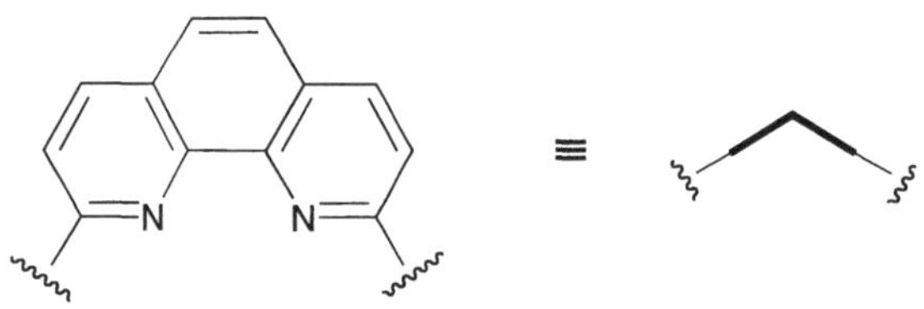

Each ring needs to make four crossings, so we need chains with four ligand units. We need not specify the details of the linking pieces.

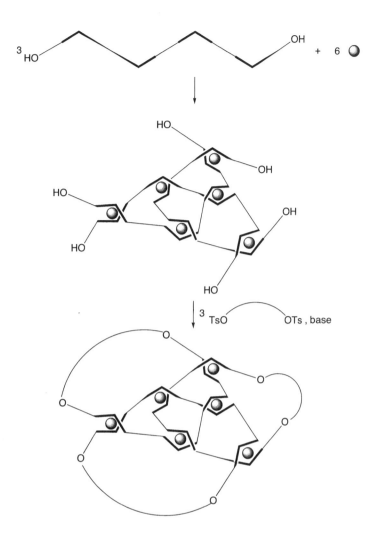

Removal of the metal atoms from the structure above will provide a Borromean ring structure.

It is probably wishful thinking to expect that the complexation of tetraligand units will actually provide the desired trimer shown unless we expend considerable effort in designing linkers that will promote this structure over the myriad of other possibilities. Such a design, to be successful, would likely require computational modeling to analyze the energy changes that come with various complexation options. The goal would be to find linkers that would give the desired supramolecular structure by thermodynamic control.

To illustrate that the strategy above is not the only one possible, we will present one alternative. Remember that topology is completely insensitive to any operations that do not break bonds — we can stretch or fold the desired ring structure any way we want. Let's pursue the strategy of making two of the rings simple and flat, leaving all the contortions for the third ring. Since no two rings are linked to each other, we can take two and pull them apart, letting the third ring stretch. The resulting structure is shown below. (It can be challenging to visualize what happens as we do this. A simple way to work it out is with a physical model, such as two key rings and a loop of string.)

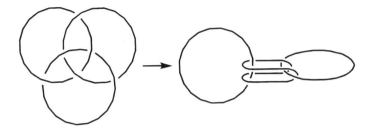

Applying the same synthetic strategy, our target structure becomes

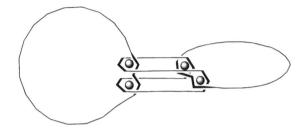

It's interesting (and fortunate) to note that this strategy requires only eight phenanthroline units and four metal atoms. A synthetic sequence to this target can make use of preformed rings:

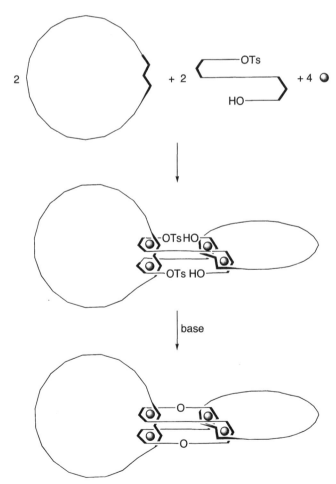

Though this strategy may also suffer from other possible complexes and linkages, it appears to be simpler than the previous strategy. The number of possibilities (and therefore, the reduction in entropy) is greatly decreased. If our linker in the TsO-linker-OH component is rigid and gives the component a 90° twist, the desired product might be preferred over alternatives. Removal of the metal atoms would give a Borromean ring structure.

The successful synthesis of a Borromean ring structure by Stoddart and coworkers used a strategy similar in some ways to the first one described above but aimed at the orthogonal geometry shown below. These researchers describe the thermodynamically driven, one-pot assembly of 18 components (12 organic components of two types plus 6 zinc ions) in essentially quantitative yield! Also worth noting is the prior construction of DNA-based Borromean rings by Seeman and coworkers (C. Mao, W. Sun, and N. C. Seeman, *Nature* **1997**, *386*, 137).

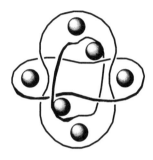

7

Energy Surfaces and Kinetic Analyses

SOLUTIONS TO EXERCISES

1. We start by writing rate laws for the formation of the two products:

$$\frac{d[P_1]}{dt} = k_1[I_1] \text{ and } \frac{d[P_2]}{dt} = k_2[I_2]$$

These rate laws will determine the ratio of products as long as the formation of products is irreversible (in other words, as long as the system is under kinetic control).

$$\frac{[P_2]}{[P_1]} = \frac{d[P_2]/dt}{d[P_1]/dt}$$

$$\frac{[P_2]}{[P_1]} = \frac{k_2[I_2]}{k_1[I_1]}$$

$$\frac{[P_2]}{[P_1]} = K_{eq}\frac{k_2}{k_1}$$

If the system is under kinetic control, the product ratio depends on the relative energies of the two transitions states, not on the relative energies of the intermediates. This is not immediately apparent from the equation we just derived, given that K_{eq} appears in the equation! We should realize that there is a cancellation effect here: if I_1 is brought higher in energy while keeping I_2 and the transition states the same, then the increase in K_{eq} is counteracted by the increase in k_1.

If the system is under thermodynamic control, then the above equation does not apply, and the product ratio is determined by the relative stability of the products. In order to achieve thermodynamic control, formation of the products (or at least the less stable one) must be reversible.

2. If the barrier that separates the two intermediates is higher that the barriers to product formation:

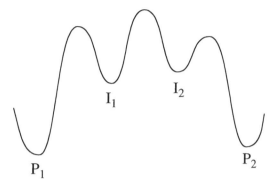

then product formation will be faster than equilibration of the intermediates. In this case, the relative energies of the intermediates will determine the product ratio, *as long as* the intermediates are able to equilibrate before the establishment of conditions that promote product formation. For example, if another reactant is necessary for product formation, then the intermediates may be free to equilibrate until this reactant is introduced. If, however, the intermediates are formed under conditions in which they can immediately react to form the products, then the product ratio will be determined by the relative rates of formation of the two intermediates, not by their relative energies.

3. The half-life is defined as the time required for half of the reactant to be consumed. Therefore, if $[A] = [A]_0$ at $t = 0$, then $[A] = [A]_0/2$ at $t = t_{1/2}$. We can simultaneously plug these quantities into the first-order integrated rate law:

$$\ln[A] = \ln[A]_0 - kt$$

$$\ln\frac{[A]_0}{2} = \ln[A]_0 - kt_{1/2}$$

$$\ln[A]_0 - \ln 2 = \ln[A]_0 - kt_{1/2}$$

$$\ln 2 = kt_{1/2}$$

$$t_{1/2} = \frac{\ln 2}{k} = \frac{0.693}{k}$$

4. The rate law for Eq. 7.27 is

$$\frac{-d[A]}{dt} = k[A]^2$$

We rearrange terms to set up the integration:

$$-\int_{[A]_0}^{[A]} \frac{d[A]}{[A]^2} = k \int_0^t dt$$

The left-hand integral fits the integral equation: $\int x^n dx = \frac{x^{n+1}}{n+1}$, where $n = -2$.

$$-\left[\frac{-1}{[A]}\right]_{[A]_0}^{[A]} = k[t]_0^t$$

$$\frac{1}{[A]} - \frac{1}{[A]_0} = kt$$

This is the integrated rate law. Plotting $\frac{1}{[A]}$ vs. t should give a straight line with slope $= k$ and

y-intercept $= \frac{1}{[A]_0}$.

The rate law for Eq. 7.28 is

$$\frac{-d[A]}{dt} = k[A][B]$$

In this case, we have two concentration variables, so we need to relate one to the other. Even if $[A]_0 \neq [B]_0$, the following will be true from the stoichiometry of Eq. 7.28:

$$[P] = [A]_0 - [A] = [B]_0 - [B]$$

We can rewrite the rate expression in terms of product formation:

$$\frac{d[P]}{dt} = k[A][B] = k([A]_0 - [P])([B]_0 - [P])$$

$$\int_0^{[P]} \frac{d[P]}{([A]_0 - [P])([B]_0 - [P])} = k \int_0^t dt$$

Looking a little farther into a table of integrals, we can find one that fits the left-hand integral:

$$\int \frac{dx}{(a+bx)(a'+b'x)} = \frac{1}{ab'-a'b}\ln\left(\frac{a'+b'x}{a+bx}\right), \text{ where } a = [A]_0, \; a' = [B]_0, \text{ and } b = b' = -1.$$

$$\frac{1}{[B]_0 - [A]_0}\left[\ln\left(\frac{[B]_0 - [P]}{[A]_0 - [P]}\right)\right]_0^{[P]} = kt$$

$$\frac{1}{[B]_0 - [A]_0}\left(\ln\left(\frac{[B]}{[A]}\right) - \ln\left(\frac{[B]_0}{[A]_0}\right)\right) = kt$$

$$\frac{1}{[B]_0 - [A]_0}\left(\ln\left(\frac{[A]_0[B]}{[B]_0[A]}\right)\right) = kt$$

If we plot the left side of the equation vs. t, we should obtain a straight line with slope = k and y-intercept = 0.

Note that this rate law will not work if $[A]_0 = [B]_0$, due to a division by zero. In this special case it will also be true at all times that $[A] = [B]$, so the rate law derived for Eq. 7.27 will work.

5. It is convenient to put the data into a table. (Most convenient is a spreadsheet, which will also allow the manipulation and plotting of the data.) For an Arrhenius plot, we need to plot $\ln(k)$ vs. $1/T$.

k (M^{-1}s^{-1})	T(K)	$\ln(k)$	$1/T$ (K^{-1})
522	592	6.258	0.001689
755	603	6.627	0.001658
1700	627	7.438	0.001595
4020	652	8.299	0.001534

We next plot the points and fit them to a line. The line shown is calculated by linear regression (a standard function in most spreadsheet programs).

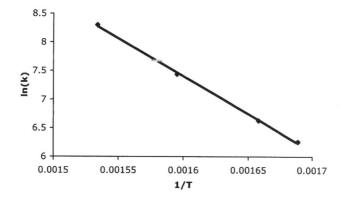

The slope of this line is $-E_a/R$, so

$$\frac{-E_a}{R} = -13130K$$

$$E_a = R(13130K) = (0.001987 \text{ kcal/mol K})(13130K) = 26.1 \text{ kcal/mol}$$

The y-intercept is ln A, so

$$\ln A = 28.42$$

$$A = 2.198 \times 10^{12}$$

$$\log A = 12.3$$

Since this value of log A is less than $(10.8 + \log T) = 13.6$ (at 600K, see Eq. 7.19), we know that $\Delta S^{\ddagger}$ is negative. This is consistent with a loss of entropy in the approach to the transition state. $\Delta S^{\ddagger}$ is usually negative for second order reactions, in which two molecules much approach each other and lose translational freedom (and usually orientational freedom) in the process of becoming the activated complex. In fact, second order reactions very often exhibit values much smaller than 12.3. Thus, one might surmise that the orientational requirements of the activated complex are not too restrictive.

The value of E_a of 26.1 kcal/mol is significantly greater than 20 kcal/mol, often taken as a benchmark value. Reactions with E_a values around 20 kcal/mol usually proceed on a convenient lab time scale (seconds or minutes) at room temperature. So the value of 26.1 kcal/mol is consistent with a reaction that requires elevated temperature.

6. Eq. 7.41 is

$$\ln\left(\frac{k_f[A]_0 - k_r[B]_0}{k_f[A]_0 - k_r[B]_0 - (k_f + k_r)[x]}\right) = (k_f + k_r)t$$

Following the procedure described in the text, we define the equilibrium extent of reaction as $[x]_e$ and set the forward and reverse rates to be equal at equilibrium:

$$k_f\left([A]_0 - [x]_e\right) = k_r\left([B]_0 + [x]_e\right)$$

Next, we rearrange terms so as to solve for an expression that appears twice in Eq. 7.41:

$$k_f[A]_0 - k_f[x]_e = k_r[B]_0 + k_r[x]_e$$

$$k_f[A]_0 - k_r[B]_0 = k_f[x]_e + k_r[x]_e = \left(k_f + k_r\right)[x]_e$$

Substituting into Eq. 7.41:

$$\ln\left(\frac{(k_f + k_r)[x]_e}{(k_f + k_r)[x]_e - (k_f + k_r)[x]}\right) = (k_f + k_r)t$$

$$\ln\left(\frac{[x]_e}{[x]_e - [x]}\right) = (k_f + k_r)t$$

7. Given: $[B]_0 = 0$ and $k_f \gg k_r$. We can then take $k_f + k_r = k_f$. Substituting into Eq. 7.41:

$$\ln\left(\frac{k_f[A]_0}{k_f[A]_0 - k_f[x]}\right) = k_f t$$

$$\ln\left(\frac{[A]_0}{[A]_0 - [x]}\right) = k_f t$$

$$\ln\frac{[A]_0}{[A]} = k_f t$$

This is just the integrated rate equation for a first order reaction. This makes sense, since the reaction of A to B is first order and we are ignoring the reverse reaction.

8. a. The primary effect of going from a 1° to 2° alkyl group is to stabilize the carbocation at the lower left corner of the diagram. This will tend to move the transition state towards this corner.
b. Since diisopropylamide is a stronger base, the entire left side of the diagram will increase in energy. This will push the transition state along the diagonal toward the reactants and perpendicular to the diagonal towards the upper right corner. Overall, the transition state will move toward the top edge (moving roughly parallel to the left side).
c. Since iodide is a better leaving group, the entire bottom edge will decrease in energy. The transition state will move parallel to the bottom edge towards the left side (the sum of the vectors pointing away from the products along the diagonal and towards the lower left corner).

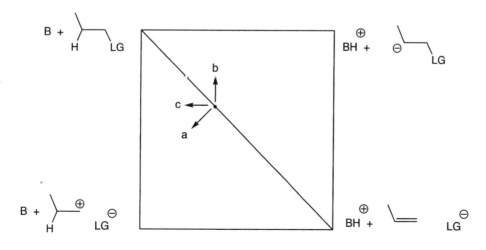

9. The Eyring equation that we use for plotting data (Eq. 7.22) shows that the enthalpy term is most influential at low temperature, since $\Delta H^{\ddagger}$ is multiplied by 1/T. The entropy term, $\Delta S^{\ddagger}$, has no such multiplier, so it will dominate at high enough temperatures. A mechanism with a favorable (low) $\Delta H^{\ddagger}$ and unfavorable (low) $\Delta S^{\ddagger}$ will therefore dominate at low temperature, while a mechanism with a favorable $\Delta S^{\ddagger}$ and unfavorable $\Delta H^{\ddagger}$ (both high) will dominate at high temperature. This will lead to an Eyring plot like the one shown below (solid lines). Note that higher temperatures are at the left in this plot, that the slope is equal to $-\Delta H^{\ddagger}/R$, and that the intercept is equal to $\Delta S^{\ddagger}/R$.

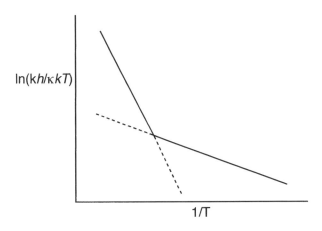

GOING DEEPER

A simple way to think about the problem posed above is to realize that if two Eyring lines (each representing a different mechanism) cross, then the one that dominates at any given temperature is the one that is higher on the vertical axis – the one with the higher rate constant. Thus, it seems that any kinked Eyring plot should look roughly like the one above, with the bend in the downward direction.

But take a look again at the Eyring plot in the highlight on vitamin B_{12} in Section 7.2.5. This plot is bent the other way, such that it appears like the dashed lines in the plot above! How can this happen? It appears that the disfavored mechanism at any given temperature is the one that dominates!

The simple analysis above makes the assumption that both mechanisms are equally possible at all temperatures, such that the relative $\Delta H^{\ddagger}$ and $\Delta S^{\ddagger}$ determine the dominant path. This is not necessarily true. The explanation offered in the highlight is that the enzyme undergoes a conformational change to an inactive form below 30°C. In other words, the reaction is less favorable at low temperatures than would be predicted from the high-temperature Eyring parameters due to a lower concentration of the catalyst. The generalization developed above may be true in simple cases with two competing mechanisms, but if other processes interfere, all bets are off!

10. The speed of light is 3.0×10^8 m/s, so the distance traveled in 10 fs is

$$(10\times10^{-15} \text{ s})(3.0\times10^8 \text{ m/s}) = 3.0\times10^{-6} \text{ m} = 3.0 \ \mu\text{m}$$

So the distance needed for the probe beam is 1.000003 m. Very high precision is required!

11. To avoid the pure trial-and-error approach of doing first- and second-order plots of the data and looking for linearity, we can get a quick sense of whether the decay is exponential just by checking for a constant half-life. Half of the initial concentration, 0.0165 M, would be 0.0083 M, corresponding to a half-life of approximately 1500 s. Halving the concentration again gives 0.0042 M, which is pretty close to the value of 0.0039 M that was observed after another 1500 s. So it appears that the half-life is reasonably constant, suggesting that we should try a first-order treatment of the data: a plot of $\ln[N_2O_5]$ vs. t.

t, s	$[N_2O_5]$	$\ln[N_2O_5]$
0	0.0165	-4.10
600	0.0124	-4.39
1200	0.0093	-4.68
1800	0.0071	-4.95
2400	0.0053	-5.24
3000	0.0039	-5.55
3600	0.0029	-5.84

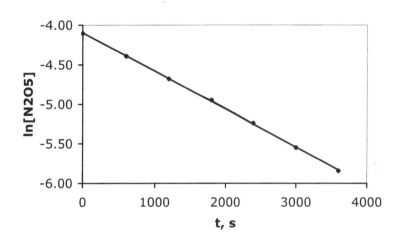

The plot is indeed linear, confirming that the decomposition follows a first-order rate law. The rate constant is just the negative of the slope, which by linear regression is $4.8\times10^{-4} \text{ s}^{-1}$.

12. The rate depends only on the catalyst concentration, not the norbornene concentration:

$$\frac{-d[\text{norbornene}]}{dt} = k[\text{catalyst}]$$

Remembering that the catalyst is not consumed in the reaction (so [catalyst] is a constant), integration is easy:

$$-\int d[\text{norbornene}] = k[\text{catalyst}]\int dt$$

$$[\text{norbornene}]_0 - [\text{norbornene}] = k[\text{catalyst}]t$$

This equation shows that the consumption of norbornene proceeds linearly with time.

13. The 1-substituted product (the left path) is favored at 80°C due to kinetic control. At this temperature, there is not enough thermal energy to allow the reverse reaction, and the sulfonation is irreversible. The major product is the one that has the lower activation free energy. The 1-substituted intermediate has two resonance structures that preserve the aromaticity of the second ring, so it lies at lower energy than the 2-substituted intermediate, which has only a single resonance structure that preserves aromaticity. This difference in intermediate energies translates to a difference in transition state energies ($\Delta\Delta G^{\ddagger}$).

The 2-substitutued product is favored at 160°C due to thermodynamic control. The reaction is reversible at this temperature, so the more stable product (by $\Delta\Delta G$) is the major one. The 1-substituted product is destabilized by a steric interaction between the sulfonate group and the H attached to C-8 (shown in the product structure). Note that this steric interaction is much less important in the intermediate, since the sulfonate group is out of the ring plane.

14. Alkyl groups *are* indeed *ortho/para* directing when the reaction is run under kinetic control, though the *ortho* isomer is generally disfavored by sterics. However, under thermodynamic control, promoted by extended reaction times and/or higher temperatures, and with sufficient EtBr, the 1,3,5-triethyl product is favored. This product allows the highest degree of alkylation without *ortho* substituents. Since the alkylation reaction is exothermic, it is thermodynamically favorable to alkylate as much as possible. The *para*-diethyl product can be converted to the 1,3,5-triethyl product through either dealkylation/alkylation or 1,2-isomerization equilibria, and both processes are known to occur under appropriate conditions. After equilibration in the absence of EtBr, the diethylbenzenes are found to be 69% *meta*, 28% *para*, and 3% *ortho*, reflecting the 2:1 statistical *meta:para* preference and the steric congestion of the *ortho* isomer.

15. The two mechanisms are dissociative and associative.

Dissociative:

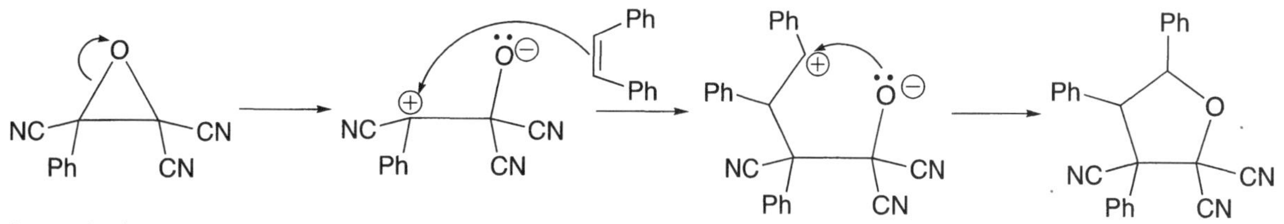

Associative:

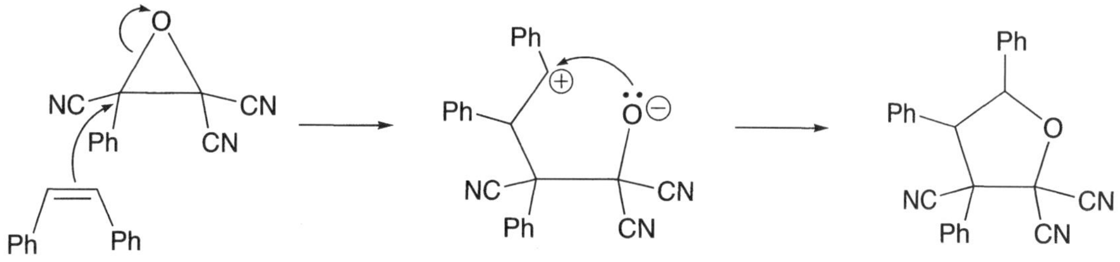

Since the first step in both reactions involve C–O cleavage to form zwitterions, and no other steps involve σ bond cleavage, it may reasonably be assumed that the first step is rate-determining in both cases. The rate laws are then derived only from the first steps:

Dissociative: rate = k[epoxide]

Associative: rate = k[epoxide][stilbene]

The rate laws differ in the effect of stilbene. Therefore, two rate measurements could distinguish the two mechanisms. For example, the initial rate could be measured with [epoxide] = [stilbene] = 0.1 M and then with [epoxide] = 0.1 M and [stilbene] = 0.2 M. If the rate is the same for the two experiments, the associative path is eliminated and the dissociative path is supported. If the rate is twice as high for the second experiment, the dissociative path is eliminated and the associative path is supported.

16. Nucleophilic catalysis:

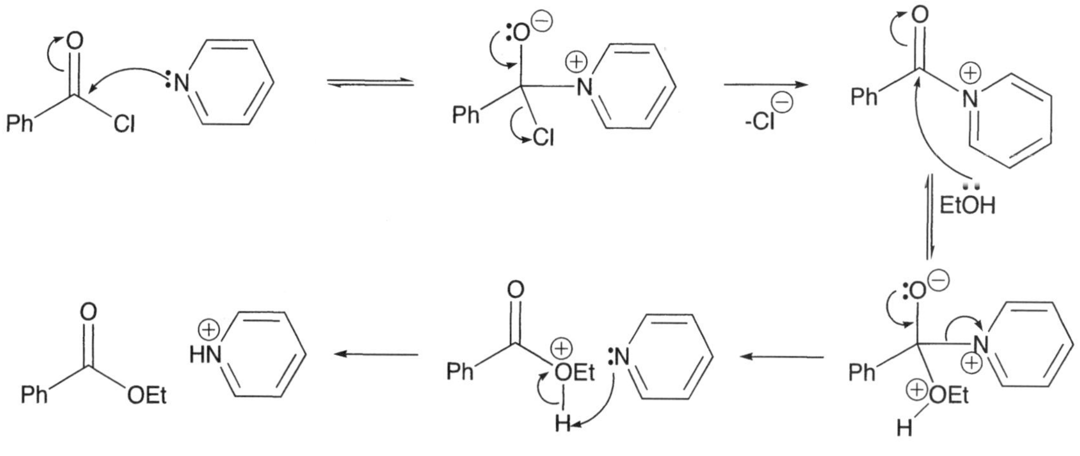

Pyridine only as base:

Since the second mechanism is simpler, we will start with that. The last, acid-base step will clearly be fast and therefore after the rate-determining step. This leaves only two steps to consider with a single intermediate for which we can take the steady-state approximation. Following the analogous example in the chapter (Eq. 7.47), we can get

$$\text{rate} = \frac{d[P]}{dt} = \frac{k_1 k_2 [\text{PhCOCl}][\text{EtOH}]}{k_{-1} + k_2}$$

In the nucleophilic catalysis mechanism, the presence of more steps and intermediates complicates the analysis, but the important point can be made easily. No matter which step is rate-determining, we can say that

$$\text{rate} \propto [\text{PhCOCl}][\text{pyridine}]$$

Thus, for this mechanism, the rate should depend on the concentration of pyridine, while is clearly should not if pyridine is involved only as a base.

17. If the acid HA is stronger, that means the entire bottom edge (where A⁻ is present) will decrease in energy. Lowering the lower left corner has the effect of moving the transition state perpendicular to the diagonal and towards this corner. Lowering the lower right corner has the effect of moving the transition state away from this corner. Overall, the transition state should move to the left and approximately parallel to the bottom edge.

Perhaps surprisingly, increasing the acid strength is predicted to have little effect on the extent of protonation at the transition state (on the diagram, the distance traveled in the down direction). In contrast, the extent of nucleophilic attack at the transition state (the distance traveled towards the right) is predicted to decrease.

The result does seem reasonable if we reflect on what the More O'Ferrall/Jencks plot shows. Our initial impression might be that increasing the acid strength should increase protonation and decrease nucleophilic attack at the transition state. This trend is represented in the diagram by the arrow pointed towards the lower left corner as this corner is lowered in energy. What we must remember is that the products (lower right corner) are also stabilized, giving a more exothermic reaction and an earlier transition state (arrow pointed toward reactants). The earlier transition state implies less protonation and less nucleophilic attack. Thus the protonation effects tend to cancel, and the nucleophilic attack effects reinforce each other.

18. What we need first is a rate law that shows first order dependencies on both reactants:

$$\text{rate} = k[\text{Ti complex}][\text{PhC}\equiv\text{CPh}]$$

But at high $[\text{PhC}\equiv\text{CPh}]$, the dependence on this reactant is lost, meaning that it must somehow be removed from the equation. This would be possible if we had $[\text{PhC}\equiv\text{CPh}]$ also in the denominator:

$$\text{rate} = \frac{k[\text{Ti complex}][\text{PhC}\equiv\text{CPh}]}{[\text{PhC}\equiv\text{CPh}]} = k[\text{Ti complex}]$$

How can we accommodate both of these behaviors in the same rate law? We can do it by setting up a competition in the denominator:

$$\text{rate} = \frac{k[\text{Ti complex}][\text{PhC}\equiv\text{CPh}]}{x + [\text{PhC}\equiv\text{CPh}]}$$

Now, if $x \gg [\text{PhC}\equiv\text{CPh}]$, we have first order dependence on $[\text{PhC}\equiv\text{CPh}]$. If $[\text{PhC}\equiv\text{CPh}] \gg x$, $[\text{PhC}\equiv\text{CPh}]$ drops out of the equation. This is perfect: $[\text{PhC}\equiv\text{CPh}]$ drops out when its value becomes large.

This is exactly the type of rate law expected for a preequilibrium mechanism (Eq. 7.47). The rate dependence on $\text{PhC}\equiv\text{CPh}$ shows a typical saturation behavior (Section 7.5.3). From these considerations, the mechanism becomes obvious:

The rate law for this mechanism is similar to what we derived above:

$$\text{rate} = \frac{k_1 k_2 [\text{Ti complex}][\text{PhC} \equiv \text{CPh}]}{k_{-1}[\text{H}_2\text{C} = \text{CMe}_2] + k_2[\text{PhC} \equiv \text{CPh}]}$$

This mechanism and rate law are also consistent with the deceleration observed with added isobutylene.

Note: There is often more than one approach to solving a problem. In fact, this problem is probably easier if one considers possible mechanisms first and derives rate laws second. For more complex mechanisms, it may be more advantageous to derive the rate law first or to take a combination approach.

19. One's quick intuition might be that entropy would be linked with temperature dependence, given that the equation $\Delta G = \Delta H - T\Delta S$ has T in the entropy term. This is true whether we are speaking in thermodynamic (ΔG) or kinetic ($\Delta G^{\ddagger}$) terms. However, we know that enthalpy (endo- or exothermicity and barrier heights) is associated with strong temperature effects. We can understand this by noting that in both the equation for equilibrium constant ($K = e^{-\Delta G/RT}$) and in the Eyring equation, division by of the ΔG term by T transfers the temperature dependence to enthalpy.

a. To see this specifically in a kinetically controlled case, we take the product ratio as the ratio of rate constants. (In the following equations, C and C' are constants.)

$$\text{product ratio} = \frac{k_1}{k_2} = \frac{CTe^{-\Delta H_1^{\ddagger}/RT} e^{\Delta S_1^{\ddagger}/R}}{CTe^{-\Delta H_2^{\ddagger}/RT} e^{\Delta S_2^{\ddagger}/R}}$$

$$= e^{\left(\Delta H_2^{\ddagger} - \Delta H_1^{\ddagger}\right)/RT} e^{\left(\Delta S_1^{\ddagger} - \Delta S_2^{\ddagger}\right)/R}$$

$$= C' e^{\left(\Delta H_2^{\ddagger} - \Delta H_1^{\ddagger}\right)/RT}$$

This shows that the temperature-dependent part of the product ratio arises from the enthalpies of activation. If the enthalpies of activation are the same, the product ratio should be temperature independent (equal to C').

b. Another possible explanation is that the product ratio is determined by dynamic effects and that transition state theory does not apply. The products formed from any given trajectory on the potential energy surface can be temperature independent, as long as energy redistribution does not occur. (Note that temperature independence of product ratio is not required for such a reaction, since the availability of different trajectories may be temperature dependent.)

20. Application of Marcus theory is relatively simple to the monoatomic ions, because there are no internal vibrations to consider, so $\lambda_i = 0$. Also, the self-exchange reactions are well defined, as they are for any electron transfer process.

21. The variables in Eq. 7.63 are $\Delta G^{\ddagger}_{int}$ and ΔG°, both of which depend on the identity of the solvent. Therefore, by using solvent-specific values for these terms, the effect of solvent is included in the calculation.

22. Butadiene has two stable conformations, both eclipsing but favored by conjugation. Cyclobutene has extremely little conformational freedom about its double bond (or any bond). Thus, the energy goes up steeply as the geometry deviates from a 0° dihedral angle.

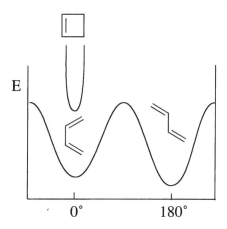

Since the product (cyclobutene) has less entropy than the reactant, ΔS for the reaction is negative. The transition state also has a restricted dihedral angle, so $\Delta S^{\ddagger}$ is also negative.

23. Since styrene decays exponentially in the presence of a large excess of bromine, we can say that the reaction is pseudo-first order and therefore first-order in styrene. The rate law clearly has a dependence on the bromine concentration, but not a first-order dependence. If it were first-order in bromine, a 1.5-fold increase in concentration should have increased the rate 1.5-fold, not 2.25-fold. We can calculate the order from the rate law:

$$\text{rate} = k[\text{styrene}][\text{Br}_2]^x$$

$$\frac{\text{rate}_2}{\text{rate}_1} = \frac{k[\text{styrene}][\text{Br}_2]_2^x}{k[\text{styrene}][\text{Br}_2]_1^x} = \left(\frac{1.5[\text{Br}_2]_1}{[\text{Br}_2]_1}\right)^x$$

$$2.25 = (1.5)^x$$

$$x = 2$$

The reaction is thus second-order in bromine. This implies that two molecules of bromine and one molecule of styrene are involved before the rate-determining step. (For a mechanism consistent with this, see Eq. 10.20.)

24. Qualitatively, the activation parameters are supportive of the mechanistic assignments. The bond cleavage to a biradical would be expected to proceed with an increase in entropy as the molecule experiences new flexibility, and this is consistent with a log A value higher than 13. In contrast, the concerted reaction should proceed through a transition state that is still as restricted as the reactant, consistent with a log A near 13.

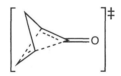

We can also suppose that a bond cleavage reaction might proceed with a higher barrier than a concerted process, in which bond formation and bond cleavage occur together. This argument is somewhat tenuous, though – we could use it to argue that all reactions should concerted! We can do better by quantitatively comparing the activation enthalpies to bond dissociation energies obtained from group increments. We do not have the increments for an acyl radical in Table 2.7, but we can do the first case. The bicyclopentane ring strain correction is taken from Figure 2.15.

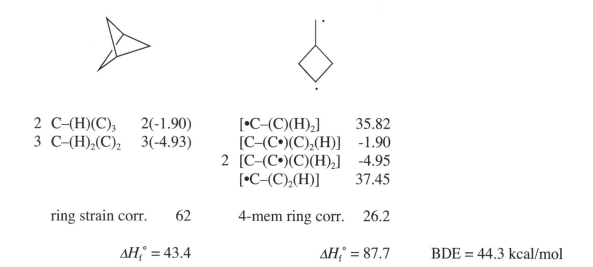

2	C–(H)(C)$_3$	2(-1.90)		[•C–(C)(H)$_2$]	35.82
3	C–(H)$_2$(C)$_2$	3(-4.93)		[C–(C•)(C)$_2$(H)]	-1.90
			2	[C–(C•)(C)(H)$_2$]	-4.95
				[•C–(C)$_2$(H)]	37.45

ring strain corr.   62   4-mem ring corr.   26.2

$\Delta H_f^\circ = 43.4$ $\Delta H_f^\circ = 87.7$ BDE = 44.3 kcal/mol

The BDE obtained is within 5 kcal/mol of the observed E_a, indicating that bond cleavage is indeed quantitatively consistent with the observed E_a. (Note that E_a represents the enthalpy difference between the reactant and transition state – not the biradical intermediate, but such a reactive intermediate is likely to be very close in energy to the transition state.) For the ketone, we might guess that the acyl group would not likely provide 20 kcal/mol of stabilization, but that the extra π bond could help to make the concerted pathway more favorable.

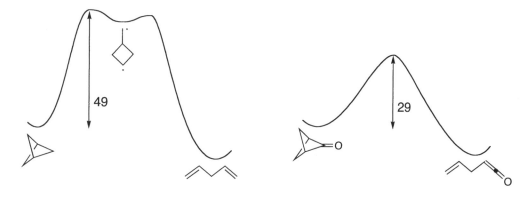

The rate constants at 120°C can be obtained by using the Arrhenius equation:

$$k = Ae^{-E_a/RT} = \left(10^{15.3}\,\text{s}^{-1}\right)e^{-(49\ \text{kcal/mol})/(0.00199\ \text{kcal/mol K})(393\text{K})} = 1.2 \times 10^{-12}\ \text{s}^{-1}$$

$$t_{1/2} = \frac{\ln 2}{k} = \frac{0.693}{1.2 \times 10^{-12}\ \text{s}^{-1}} = 5.8 \times 10^{11}\ \text{s} = 18,000\ \text{years for bicyclopentane}$$

$$k = Ae^{-E_a/RT} = \left(10^{12.9}\,\text{s}^{-1}\right)10^{12.9}e^{-(29\ \text{kcal/mol})/(0.00199\ \text{kcal/mol K})(393\text{K})} = 6.3 \times 10^{-4}\ \text{s}^{-1}$$

$$t_{1/2} = \frac{\ln 2}{k} = \frac{0.693}{6.3 \times 10^{-4}\ \text{s}^{-1}} = 1,100\ \text{s} = 18\ \text{minutes for bicyclopentanone}$$

25. The highlight says that the pK_a of the ketyl radical is determined by measuring relative amounts of the ketyl radical and the ketyl anion in solutions of different pH. A potential problem is that the pulse radiolysis (injection of electrons) could have a significant tendency to increase the solution pH. To avoid this problem, buffered solutions of acetophenone should be used. As long as the buffer capacity is high enough, the pH should not be significantly changed by the radiolysis.

We can rationalize the fact that the ketyl radical from acetophenone (pK_a = 10.5) is a stronger acid than typical alcohols by noting that the conjugate base can delocalize the negative charge. Though the best resonance form has the charge on O and the radical on C, the form with these switched is also a reasonable contributor. Further delocalization into the phenyl ring is also possible.

26. A simple qualitative result from Eq. 7.64, given that $\Delta G^{\ddagger}_{int}$ is always positive, is that $x^{\ddagger}$ will be less than 1/2 if ΔG° is negative and greater than 1/2 if ΔG° is positive.

$$x^{\ddagger} = \frac{1}{2} + \frac{\Delta G^{\circ}}{8\Delta G^{\ddagger}_{int}} \quad \text{(Eq. 7.64)}$$

This is perfectly consistent with the Hammond postulate in that exergonic reactions will have early transition states and endergonic reactions will have late transition states.

Cases where the equation and the Hammond postulate would not apply well would be reactions with relatively flat sections of the energy surface. In the Marcus analysis, this is equivalent to having parabolas with very different widths. In the following exergonic reactions, late transition states are the result of relative flat surfaces near the reactant.

8

Experiments Related to Thermodynamics and Kinetics

1. We can calculate the ratio of rate constants by using the Arrhenius equation:

$$k = Ae^{-E_a/RT}$$

$$\frac{k_H}{k_D} = \frac{Ae^{-E_a(H)/RT}}{Ae^{-E_a(D)/RT}}$$

We can take the frequency factor, A, to be the same for the two isotopes, so this will cancel out from the equation.

$$\frac{k_H}{k_D} = e^{(E_a(D)-E_a(H))/RT}$$

As illustrated in Figure 8.1,

$$E_a(D) = E_a(H) + \left(ZPE(H) - ZPE(D)\right)$$

Solving for $E_a(D) - E_a(H)$ and plugging this in, we get

$$\frac{k_H}{k_D} = e^{(ZPE(H)-ZPE(D))/RT}$$

We can calculate $ZPE(H)$ by using Eq. 8.1:

$$ZPE(\mathrm{H}) = e_0 = \frac{1}{2}h\nu(\mathrm{H}) = \frac{1}{2}hc\bar{\nu}(\mathrm{H})$$

We can plug this expression, along with the corresponding one for $ZPE(\mathrm{D})$, into the isotope effect equation:

$$\frac{k_\mathrm{H}}{k_\mathrm{D}} = e^{\left(\frac{1}{2}hc\bar{\nu}(\mathrm{H}) - \frac{1}{2}hc\bar{\nu}(\mathrm{D})\right)/RT} = e^{hc(\bar{\nu}(\mathrm{H}) - \bar{\nu}(\mathrm{D}))/2RT}$$

Recognizing that $k = R/N_A$, we can see that this is very close to the desired equation. We need Avogadro's number, N_A, in the numerator of the exponential in order to make the desired substitution, and we can just multiply the numerator by N_A. We can do this because N_A, which equals 6.02×10^{23}/mol, is just a conversion factor and has a value of 1 (because 1 mol = 6.02×10^{23}). In effect, we are just converting the energy difference to a molar quantity.

$$\frac{k_\mathrm{H}}{k_\mathrm{D}} = e^{hcN_A(\bar{\nu}(\mathrm{H}) - \bar{\nu}(\mathrm{D}))/2RT} = e^{hc(\bar{\nu}(\mathrm{H}) - \bar{\nu}(\mathrm{D}))/2kT}$$

2. We already have everything we need to plug into the equation from exercise 1 except for $\bar{\nu}(\mathrm{D}) = \bar{\nu}(\mathrm{C-D})$, which we can calculate from the corresponding $\bar{\nu}(\mathrm{H}) = \bar{\nu}(\mathrm{C-H}) = 3000$ cm^{-1} by adjusting for the change in reduced mass. Applying Eq. 8.2,

$$\bar{\nu}(\mathrm{C-D}) = \frac{\nu(\mathrm{C-D})}{c} = \frac{1}{2\pi c}\sqrt{\frac{k}{m_r(\mathrm{C-D})}}$$

The force constant k is the same for both C–H and C–D. So we can get k from the same equation for $\bar{\nu}(\mathrm{C-H})$:

$$\bar{\nu}(\mathrm{C-H}) = \frac{1}{2\pi c}\sqrt{\frac{k}{m_r(\mathrm{C-H})}}$$

$$k = (m_r(\mathrm{C-H}))(2\pi c\bar{\nu}(\mathrm{C-H}))^2$$

Plugging this into the equation for $\bar{\nu}(\mathrm{C-D})$ gives

$$\bar{\nu}(\mathrm{C-D}) = \frac{1}{2\pi c}\sqrt{\frac{(m_r(\mathrm{C-H}))(2\pi c\bar{\nu}(\mathrm{C-H}))^2}{m_r(\mathrm{C-D})}}$$

$$= \bar{\nu}(\mathrm{C-H})\sqrt{\frac{m_r(\mathrm{C-H})}{m_r(\mathrm{C-D})}}$$

$$= 3000 \text{ cm}^{-1} \sqrt{\frac{(12 \cdot 1)/(12+1)}{(12 \cdot 2)/(12+2)}}$$

$$= 2200 \text{ cm}^{-1}$$

We are now ready to plug values into the equation from exercise 1.

$$\frac{k_H}{k_D} = e^{hc(\bar{v}(H)-\bar{v}(D))/2kT}$$

$$= e^{\left(6.63 \times 10^{-34} \text{ J·s}\right)\left(3.00 \times 10^{10} \text{ cm/s}\right)\left(3000 \text{ cm}^{-1}-2200 \text{ cm}^{-1}\right)/2\left(1.38 \times 10^{-23} \text{ J/K}\right)(298\text{K})}$$

$$= e^{1.93} = 6.9$$

3. The strategy for this secondary KIE is the similar to that for exercise 2, except that we will need to calculate *ZPEs* for both the reactant and transition state. Referring to the diagram below, we can relate the activation energies to the appropriate *ZPEs* and plug them into the equation from exercise 1.

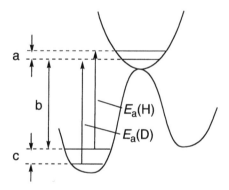

$$\frac{k_H}{k_D} = e^{(E_a(D)-E_a(H))/RT} = e^{[(b+c)-(a+b)]/RT} = e^{(c-a)/RT} = e^{\left((ZPE(H)-ZPE(D))-\left(ZPE(H)^{\ddagger}-ZPE(D)^{\ddagger}\right)\right)/RT}$$

$$\frac{k_H}{k_D} = e^{\left(\left(\frac{1}{2}hc\bar{v}(H)-\frac{1}{2}hc\bar{v}(D)\right)-\left(\frac{1}{2}hc\bar{v}(H)^{\ddagger}-\frac{1}{2}hc\bar{v}(D)^{\ddagger}\right)\right)/RT} = e^{hc\left((\bar{v}(H)-\bar{v}(D))-\left(\bar{v}(H)^{\ddagger}-\bar{v}(D)^{\ddagger}\right)\right)/2RT}$$

$$\frac{k_H}{k_D} = e^{hc\left((v(H)-v(D))-\left(v(H)^{\ddagger}-v(D)^{\ddagger}\right)\right)/2kT}$$

The maximum effect will occur if the rehybridization from sp^3 to sp^2 is complete at the transition state. Thus, we can take the relevant frequencies as 1350 and 800 cm^{-1} for the H reactant and transition state, respectively, as given in Figure 8.5.

To get the frequencies for D, we use the relationship derived from Eq. 8.2 in exercise 2, assuming that it applies equally well to bending vibrations.

$$\overline{v}(C-D) = \overline{v}(C-H)\sqrt{\frac{m_r(C-H)}{m_r(C-D)}} = \overline{v}(C-H)\sqrt{\frac{(12 \bullet 1)/(12+1)}{(12 \bullet 2)/(12+2)}} = 0.734\,\overline{v}(C-H)$$

$$= 0.734\left(1350 \text{ cm}^{-1}\right) = 991 \text{ cm}^{-1}$$

$$\overline{v}(C-D)^{\ddagger} = 0.734\left(800 \text{ cm}^{-1}\right) = 587 \text{ cm}^{-1}$$

Plugging in all the values, we get

$$\frac{k_H}{k_D} = e^{\left(6.63\times10^{-34} \text{ J·s}\right)\left(3.00\times10^{10} \text{ cm/s}\right)\left(\left(1350 \text{ cm}^{-1}-991 \text{ cm}^{-1}\right)-\left(800 \text{ cm}^{-1}-587 \text{ cm}^{-1}\right)\right)/2\left(1.38\times10^{-23} \text{ J/K}\right)(298\text{K})}$$

$$= e^{0.353} = 1.4$$

4. Assuming two products and two reactants, each with one exchangeable hydrogen, we can start with the equilibrium expressions:

$$K_H = \frac{[P_1-H][P_2-H]}{[R_1-H][R_2-H]} \text{ and } K_D = \frac{[P_1-D][P_2-D]}{[R_1-D][R_2-D]}$$

$$\frac{K_D}{K_H} = \frac{[P_1-D][P_2-D]}{[R_1-D][R_2-D]} \Bigg/ \frac{[P_1-H][P_2-H]}{[R_1-H][R_2-H]}$$

Rearranging factors gives

$$\frac{K_D}{K_H} = \left(\frac{[P_1-D]}{[P_1-H]}\right)\left(\frac{[P_2-D]}{[P_2-H]}\right) \Bigg/ \left(\frac{[R_1-D]}{[R_1-H]}\right)\left(\frac{[R_2-D]}{[R_2-H]}\right)$$

We can now bring in solvent concentrations, multiplying numerator and denominator by the same quantity:

$$\frac{K_D}{K_H} = \left(\frac{[P_1-D]}{[P_1-H]}\right)\left(\frac{[P_2-D]}{[P_2-H]}\right)\left(\frac{[S-H][S-H]}{[S-D][S-D]}\right) \Bigg/ \left(\frac{[R_1-D]}{[R_1-H]}\right)\left(\frac{[R_2-D]}{[R_2-H]}\right)\left(\frac{[S-H][S-H]}{[S-D][S-D]}\right)$$

$$\frac{K_D}{K_H} = \left(\frac{[P_1-D][S-H]}{[P_1-H][S-D]}\right)\left(\frac{[P_2-D][S-H]}{[P_2-H][S-D]}\right) \Bigg/ \left(\frac{[R_1-D][S-H]}{[R_1-H][S-D]}\right)\left(\frac{[R_2-D][S-H]}{[R_2-H][S-D]}\right)$$

$$\frac{K_D}{K_H} = \frac{\phi_1^p \phi_2^p}{\phi_1^r \phi_2^r}$$

Within the limitations imposed, i and $j = 1$ or 2, we have derived Eq. 8.11:

$$\frac{K_D}{K_H} = \frac{\prod_i \phi_i^p}{\prod_j \phi_j^r}$$

5. We start with the Eyring equation, where "C" is a constant and a "0" subscript represents the reference substituent:

$$k_0 = CTe^{-\Delta G_0^{\ddagger}/RT}$$

$$\Delta G_0^{\ddagger} = -RT\ln\left(\frac{k_0}{CT}\right)$$

$$\Delta G_x^{\ddagger} - \Delta G_0^{\ddagger} = -2.303RT\left(\log\left(\frac{k_x}{CT}\right) - \log\left(\frac{k_0}{CT}\right)\right) = -2.303RT\log\left(\left(\frac{k_x}{CT}\right)\left(\frac{CT}{k_0}\right)\right) = -2.303RT\log\left(\frac{k_x}{k_0}\right)$$

We define the substituent constant: $C_x = \log\left(\frac{k_x}{k_0}\right)$

$$\Delta G_x^{\ddagger} - \Delta G_0^{\ddagger} = -2.303RT(C_x)$$

The above equation can be taken as applying to the reference reaction. We can write the same equation for a different reaction, using primes to distinguish the two:

$$\Delta G_x^{\ddagger'} - \Delta G_0^{\ddagger'} = -2.303RT\left(C_x'\right)$$

We can now define the reaction constant, Q:

$$Q = \frac{C_x'}{C_x} = \frac{\left(\Delta G_x^{\ddagger'} - \Delta G_0^{\ddagger'}\right)}{\left(\Delta G_x^{\ddagger} - \Delta G_0^{\ddagger}\right)} = \frac{\Delta\Delta G^{\ddagger'}}{\Delta\Delta G^{\ddagger}}$$

$$C_x' = QC_x$$

$$\log\left(\frac{k_x'}{k_0'}\right) = QC_x$$

This equation was derived for a general case. To convert it specifically to the Hammett equation, we can take $Q = \rho$ and $C_x = \sigma_x$. We can also drop the primes and specify the reference substituent as H.

$$\log\left(\frac{k_x}{k_H}\right) = \rho\sigma_x$$

6. a. Section 8.6.2 lists three situations in which a linear free energy relationship might be obtained. The first two require coincidentally equal $\Delta\Delta H^{\ddagger}$ or $\Delta\Delta S^{\ddagger}$ values for the reference reaction and reaction of interest. (The text discussion made use of equilibrium values – $\Delta\Delta H$ and $\Delta\Delta S$ – but the same arguments apply equally well to kinetic values.) In general, this is unlikely, since the steric effects of substituents in different reactions are likely to be different, especially given that the distance of the substituents from the site of reaction might be different. The third possibility is the most likely: that $\Delta\Delta H^{\ddagger}$ and $\Delta\Delta S^{\ddagger}$ are linearly related for both reactions, as they would be if enthalpy and entropy compensate for each other.

As described in Section 8.6.4, steric effects can lead to enthalpy/entropy compensation. For example, as the steric bulk of the reactants in a bimolecular reaction increases, the transition state might occur earlier due to increased steric interference. In this case, the new bonding interactions would be weaker and the reactants would retain more of their entropy at the transition state. Both $\Delta H^{\ddagger}$ and $\Delta S^{\ddagger}$ would increase (become more positive or less negative). If these effects scale linearly, a linear free energy relationship would be obtained.

Note that the third possibility of linearly related $\Delta\Delta H^{\ddagger}$ and $\Delta\Delta S^{\ddagger}$ values do not require that enthalpy and entropy compensate for each other. A linear relationship could also be obtained if $\Delta H^{\ddagger}$ and $\Delta S^{\ddagger}$ went in opposite directions, augmenting each other.

b. Grunwald Winstein LFERs deal with effects of solvent polarity, for which enthalpy/entropy compensation is commonly observed. Solvation of charges or partial charges is enthalpically favorable, since some of the charge is effectively delocalized into the solvent. However, this solvation occurs through the ordering of solvent molecules such that their negative ends point toward a positive charge, for example. The more ordered arrangement of solvent molecules is associated with a decrease in entropy. Thus, the lower enthalpy is compensated to some extent by a lower entropy. If charge is created on going from the reactants to the transition state, increases in solvent polarity will lead to lower $\Delta H^{\ddagger}$ and more negative $\Delta S^{\ddagger}$.

7. The half-life for a unimolecular reaction is given by

$$t_{1/2} = \frac{\ln 2}{k}$$

where k can be calculated by the Arrhenius equation:

$$t_{1/2} = \frac{\ln 2}{Ae^{-E_a/RT}} = \frac{\ln 2}{A}e^{E_a/RT}$$

If $A = 10^{13}$ s^{-1} and $E_a = 0.5$ kcal/mol, then

$$t_{1/2} = \left(\frac{\ln 2}{10^{13} \text{ s}^{-1}}\right)e^{(0.5 \text{ kcal/mol})/(0.00199 \text{ kcal/mol} \cdot \text{K})T} = \left(6.93 \times 10^{-14} \text{ s}\right)e^{251\text{K}/T}$$

Plugging $T = 4K$ into this equation gives $t_{1/2} = 1.2 \times 10^{14}$ s, roughly 4 million years, and plugging in $T = 8K$ gives $t_{1/2} = 2.9$ s! The enormous change in half-life with only 4 degrees comes about partly because the temperature is actually doubled. A comparable temperature change from ambient temperature (298K) would require an increase to about 600K. But that is not the whole story. Also important is the fact that the exponent ($251 K/T$) is large in the very low temperature range. Indeed, if we double the temperature again to 16K, the change in half-life is already less dramatic: $t_{1/2} = 4.5 \times 10^{-7}$ s, a change of only seven orders of magnitude. Another factor of two in T reduces $t_{1/2}$ by a factor of less than 3,000. This makes perfect sense if we remember what a plot of $y = e^x$ looks like (shown below for two x ranges). Clearly, changing x by a factor of two has a much bigger effect on y when x is large (such as going from 40 to 80 on the right plot) than it does when x is small (such as going from 2 to 4 on the left plot).

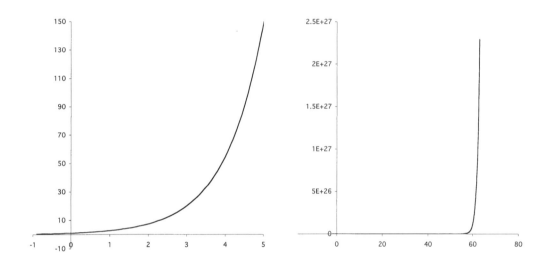

GOING DEEPER

The observation of exponential decay is actually very rare for unimolecular reactions occurring near 4K. Why is this? It is generally impossible to study reactions in liquid solution at these low temperatures, since helium is the only substance that is liquid in this range and helium is a very poor solvent. Therefore, kinetic studies in this temperature range generally employ matrix isolation, either with frozen solvents or inert gases such as argon or xenon, which are deposited as solids from the gas phase. Whether the solid matrix around a reactant precursor is formed by freezing or deposition, the exact placement of matrix molecules around the precursor and even the conformation of the precursor itself will vary from one site to the next in the matrix. When the precursor is then converted to the reactant of interest, these differences in geometry will lead to differences in activation parameters for the reaction being studied. As we can appreciate from our above analysis, even very small differences in activation energy may lead to very large differences in rate. This phenomenon is known as the matrix site effect, and the kinetics that result are called distribution kinetics, since there is a distribution of rate constants. Though the rate behavior is complicated, several strategies are useful for the analysis of such kinetics, and meaningful results can still be obtained. In some cases, Arrhenius behavior is even still observed, based for example on the most probable rate constant. Another complication is that tunneling is often observed in this temperature range, since tunneling rates are much less sensitive to temperature than rates for activated processes. See, for example, "Matrix-Isolation Decay Kinetics of Triplet Cyclobutanediyls. Observation of Both Arrhenius Behavior and Heavy-atom Tunneling in C-C Bond-forming Reactions." Sponsler, M. B.; Jain, R.; Coms, F. D.; Dougherty, D. A., *J. Am. Chem. Soc.* **1989**, *111*, 2240.

8. The temperature of interstellar space is very low. The low temperature tremendously affects the rates of activated processes, but has less of an effect on the rates of tunneling processes. Therefore, tunneling processes are often favored at very low temperature.

9. The two potential energy diagrams below illustrate the different situations described in the highlight and in the exercise.

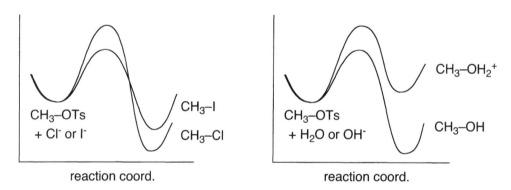

In the comparison of chloride and iodide, the better nucleophile (I⁻) leads to the weaker bond (C–I), such that the same group is also the better leaving group. In the comparison of water and hydroxide, however, the better nucleophile (OH⁻) forms a much stronger bond, so in this case the better nucleophile is the poorer leaving group. From these diagrams, it is clear that both nucleophilicity and leaving group ability depend on the energy of the transition state, but only leaving group ability depends on the strength of the bond between the group of interest and the substrate. (Actually, this bond strength does also affect nucleophilicity, but only to the extent that the bond is formed at the transition state.)

10. The Grunwald-Winstein analysis serves to elucidate the effects of solvent on a reaction. If a reaction is largely driven by release of ring strain or steric strain, it is likely to be less sensitive to changes in solvent, leading to small *m* values. In other words, strain in the reactant that is relieved in the transition state will reduce the activation energy for the reaction, reducing also the need for solvent assistance in the reaction.

11. Though an aryl sulfone is likely better at stabilizing an adjacent negative charge than is an aryl group, the resonance in the sulfone involves only the oxygen atoms on sulfur and not the aromatic ring. Therefore, the effect of a *para* substituent should be much smaller than for a

directly attached aryl group, which can accept the adjacent charge into the ring through resonance.

12. The base-catalyzed hydrolysis is easy to analyze, since the first step is rate-determining.

In this step, the carbonyl C, with a partial positive charge, goes to a tetrahedral C with an adjacent negative charge. Through induction, some of this charge will reside on the C. So this C is going from partially positive to partially negative, a process that would be accelerated by *para* electron withdrawing substituents. This is consistent with a sizable, positive value for ρ of 2.19. Since neither is a full charge, the value is less than the 4-5 that might be observed in such a case.

In the acid-catalyzed hydrolysis, the rate-determining step is also normally the nucleophilic attack, which now occurs in the second step.

In the rate-determining step, a very significant positive charge at the carbonyl C (through resonance) becomes a much weaker positive charge (through induction). A weakening of the positive charge is the same direction as conversion of a positive to a negative, as observed for the base-catalyzed reaction, so this is consistent with a positive ρ value, as is observed. However, the observed value, 0.14, seems quite small for the very significant reduction in charge. A competing effect is that the initial equilibrium will be opposed by electron-withdrawing groups, reducing the concentration of the protonated ester, tending to decelerate the hydrolysis. On balance, expected negative ρ value for the equilibrium step partially cancels the positive value expected for the second step, such that the overall value is small and positive.

13. A more accurate statement is, "Deuterium prefers the position with the largest force constants." These force constants include not only the stretching force constant that is associated with the strength of the bond, but also bending force constants.

14. When a substitution reaction is done on a monosubstituted benzene, *para*, *meta*, and *ortho* products are possible.

The regioisomer that dominates depends on the electron donating or withdrawing characteristics of X, the same property that is most important in determining the Hammett plot. In collecting data for the Hammett plot, one must determine the absolute rate constants for formation of the isomer of interest (typically *para* or *meta*, since steric effects play a significant role in the *ortho* case). Thus, the analysis method must allow measurement of the concentration of one isomer in the presence of either larger or smaller amounts of the other isomers. While the most direct option is to follow the growth of the desired product isomer, another option is to follow the decay of the reactant and then analyze the product mixture to determine the isomer ratio. For example, if the desired isomer is found to constitute 10% of the product mixture, the rate constant for its formation can be taken as 10% of the rate constant for the disappearance of the reactant. (This method carries the assumption that the product ratio remains constant through the reaction, which should be the case as long as the products – including the HBr – do not participate in the reaction.)

For both *para* and *meta* isomers, negative ρ values are expected, due to the formation of a positive charge in the rate-determining step (the addition step). Thus, the reaction should be faster for electron-donating X than for electron-withdrawing X. The *para* ρ value should have a higher magnitude, since the charge can be delocalized to the C that is attached to X.

15. The S_N1 and S_N2 mechanisms both proceed from the same intermediate (the protonated reactant):

A simple way to distinguish these pathways is through isotope labeling. In the S_N1 mechanism, the carbonyl oxygen of the product comes from the water (the water molecule shown acting as a nucleophile in the second step), while in the S_N2 mechanism, the carbonyl oxygen originates from the orthoester (the leaving group O of the S_N2 step). Therefore, the use of $H_2^{18}O$ should give ^{18}O-labeled product if the mechanism is S_N1 and unlabeled product if the mechanism is S_N2. If labeled product is observed, one would need to verify through a control experiment that it does not come from exchange of label into the product. (Exchange is expected, but if it is shown to be significantly slower than the formation and analysis of the product under the conditions of the hydrolysis, it can be disregarded.) An alternative labeling experiment would be the use of ^{18}O-labeled orthoester.

16. a. A primary carbocation, being much less stable, would depend much more heavily on the solvent for stabilizing solvation interactions. Thus, changes in solvent polarity should have larger effects on the rate of the reaction that forms a primary carbocation.

b. The statement of part (a) does not apply to the solvolysis of ethyl tosylate. The reactions involve nucleophilic assistance of the solvent – S_N2 pathways – rather than formation of ethyl cation through an S_N1 pathway.

17. a. A quick answer is that the equilibrium constant is probably larger for the deuterated complex, since D favors the site with higher force constants, which most often goes along with stronger bonding. The more careful analysis given in part b gives the same answer.

b. The reaction coordinate for this isomerization can be taken as the H–Os–Os angle (marked on the left structure below), which represents a bending vibrational mode for the terminal H's and a mixed bending-stretching mode for the bridging H's.

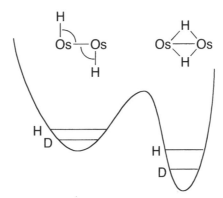

The diagram reflects both the lower energy of the bridged isomer, inferred from the equilibrium data, and a narrower well for the bridged isomer. The latter is expected because the H–Os–Os motion for the bridging H is much less free than for the terminal H, since this motion also stretches and compresses the other Os–H bond. The narrower well results in a larger force constant and a greater separation between the H and D zero-point energies. This provides for a more exothermic reaction for D relative to H, giving a larger equilibrium constant for D, as predicted also in part a.

18. Like the benzoic acid reference reaction, both of these reactions are acid base equilibria. In both cases, the reactant is neutral and the product is charged. The first reaction, with a negatively charged product, should have a positive ρ value, just like the ionization of benzoic acid. The product of the second reaction is positively charged, so in this case a negative ρ value is expected. In the enolate product of the first reaction, one of the two resonance

structures places the negative charge on the C bearing the X group. Since resonance in benzoate does not bring the charge to the C bearing the X group (or even into the aromatic ring at all), the enolization reaction shown should have a ρ value much greater than 1. (Due to the strong resonance effect, the inductively based σ values might not produce a linear plot for the enolization and σ⁻ values might give a better result.)

The charge in the protonated ketone product of the second reaction is delocalized by resonance into the aromatic ring and even to the aromatic C that bears the substituent. However, a CH₂ group separates the charge and the X group, limiting the effect of X to an inductive effect. This is a similar situation as in the benzoates, where the donating or withdrawing nature of X can be transmitted by resonance to the C bearing the carboxylate. Since the latter, however, involves interactions through a conjugated (or cross-conjugated) π system, this may be expected to produce larger substituent effects than the case that requires induction through two σ bonds. Therefore, the ρ value for protonation of these ketones will likely be between 0 and –1.

19. a. The nucleophilic attack of azide ion does not involve C–H(D) cleavage, so any KIE must be secondary. Two mechanisms might be responsible for a secondary KIE. The hybridization of the C attached to H or D goes from sp² to sp³, so an inverse KIE is expected. However, a negative charge is produced β to the C–H(D) bond, so a hyperconjugative effect is also possible (shown below). This serves to weaken the C–H(D) bond, corresponding to a normal KIE. So some cancellation of effects is expected, but which should dominate? The hyperconjugation effect should be small in this case for three reasons. First, negative hyperconjugation is less favorable than positive hyperconjugation, due to the lower electronegativity of H relative to C. Second, the negative charge will reside mostly on the O of the enolate, rather than on C. Third, at the transition state for addition of N₃⁻, the H atom should be positioned roughly in the plane, such that overlap between the C–H(D) bond and the π orbitals should be poor. Therefore, an inverse secondary KIE is expected.

b. In this case, the C–H(D) bond is not cleaved, and no charges are formed. A normal, secondary KIE is expected, due to the change in hybridization of the C attached to H or D from sp^3 to sp^2.

c. This reaction involves cleavage of the C–H(D) bond, so a normal, primary KIE is expected.

20. The first step in ozonolysis is the formation of a primary ozonide:

The observation of inverse KIEs with D substitution on either end of the π bond is consistent with this first step being the rate-determining step, as both alkene C's are changed from sp^2 to sp^3. The fact that the two KIEs have the same magnitude suggests that the transition state is symmetrical, with no greater bond formation to either the substituted or unsubstituted ends of the alkene.

21. Cross-over: use a mixture of unlabeled and doubly labeled reactant:

The formation of significant amounts of the cross-over products would be consistent with path 2, while the lack of cross-over products would be consistent with path 1 – as well as with path 2, as long as recombination of the carbanion and aldehyde is faster than escape from the solvent cage.

Trapping: attempt trapping of the benzaldehyde. The only intermediates that are not common to both mechanisms are the carbanion R$_3$C$^-$ and benzaldehyde, PhCHO. Trapping the carbanion is problematic, since two other carbanions are involved in the reaction: nBuLi and the first intermediate. A simple way to attempt trapping of the benzaldehyde would be to do the reaction with excess nBuLi. This could be done by slowly adding the ether reactant to a solution of nBuLi instead of the opposite. The trapping product would be 1-phenyl-1-pentanol. Detection or isolation of a significant yield of this product would be consistent with path 2. As with the cross-over experiment, the lack of this product would be consistent with path 1 or with path 2, as long as cage recombination is faster than escape.

Stereochemistry: use a chiral and non-racemic reactant with CRR'R". Path 1 would be likely to occur stereoselectively (with either mostly retention or inversion), while path 2 would be expected to occur with racemization. In order for path 2 to occur stereoselectively, recombination of the carbanion and benzaldehyde would have to be faster than inversion and rotation of the carbanion.

A final comment: The most widely accepted mechanism for Wittig rearrangements is neither of the two shown. It is a variation on path 2, where the fragments are a ketyl radical anion and a free radical, resulting from a homolytic cleavage rather than a heterolytic cleavage of the carbanion. This path might be distinguished from path 2 by trapping experiments, given the different reactivities of the intermediates.

22. Isotope labeling: use C_6D_5Cl. The top pathway (addition-elimination) would give $C_6D_5NH_2$, while the bottom pathway (elimination-addition) would give $o\text{-}C_6HD_4NH_2$.

KIE: use C_6H_5Cl and C_6D_5Cl. The top pathway would give only a secondary KIE, while the bottom pathway would give a primary KIE.

Trapping: attempt to trap benzyne. A trapping agent, such as anthracene, could be added to the reaction mixture in order to trap benzyne in a Diels-Alder adduct. (Note that if Birch reduction of the anthracene interferes, a different trap would be necessary.) If the adduct is not observed, the intermediacy of benzyne is not eliminated from possibility, since the reaction to give aniline might be faster than trapping. More reactive traps might help, but again the negative result is inconclusive, unless it can be shown that benzyne should be trapped under these conditions.

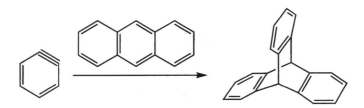

LFER: use $p\text{-}XC_6H_4Cl$. The top pathway, with a negative charge that can delocalize by resonance to the C bearing the X group, should give a linear Hammett plot with σ^- with a large, positive ρ. The bottom path, proceeding through an in-plane elimination and a neutral intermediate, should exhibit relatively small substituent effects.

23. Plotting $\Delta H^{\ddagger}$ vs. $\Delta S^{\ddagger}$, as suggested by Eq. 8.62, does indeed reveal a linear correlation. The isokinetic temperature, β, is the slope of this plot: 765 K. (Remember that eu = cal/mol K.)

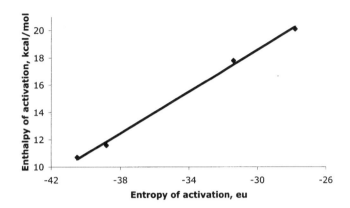

24. If R (the group trans to the oxime OH) is made to be substituted benzyl groups (p-XC$_6$H$_5$CH$_2$), a Hammett plot for the rearrangement, using σ^+ should give a large, negative ρ value if path A is operating. On the other hand, in path B, this R group migrates with its electrons, presumably with little or no positive charge generation on R. Therefore, a small negative or positive ρ value would be expected for this pathway.

In addition, if R' (the group cis to the oxime OH) is made to be substituted phenyl groups (p-XC$_6$H$_5$), a Hammett plot with σ^+ should also be informative. Path A should exhibit a small ρ value, while path B should exhibit a large, negative value.

25. a. The p-anisyl group fairly effectively delocalizes the positive charge by resonance, such that a much smaller partial charge still resides on the bicyclic ring system. With the much more stabilized and delocalized charge, the alkene π bond plays a much diminished, though still significant, role in further delocalizing the charge.

b. As noted in part a, the aryl group and the alkene both serve to delocalize the positive charge. If the alkene π electrons are not present, then the aryl group takes on a larger role in this delocalization, and this is shown by the ρ values given. A much larger value, -5.17, is observed in the absence of the alkene than in its presence, -2.30.

c. By making the X group less electron-donating or even electron-withdrawing, the aryl group becomes much less able to delocalize and stabilize the positive charge. Therefore, the alkene, when present, plays a correspondingly greater role in this regard. The π/no π relative rates increase, becoming more like the parent system (without the aryl group).

d. The Rule of Increasing Electron Demand states that a neighboring group that can serve to delocalize charge in the transition state and thus accelerate a reaction will have the largest effect when the charge has no other means for delocalization.

26. We make use of Eq. 8.18, truncated to two factors:

$$\frac{k_n}{k_H} = (1 - n + n\phi_1)(1 - n + n\phi_2)$$

In this equation, ϕ_1 and ϕ_2 are the reciprocals of the isotope effects. In the first case,

$$\phi_1 = \phi_2 = 1/2$$

$$\frac{k_n}{k_H} = (1 - n + n/2)(1 - n + n/2) = (1 - n/2)^2 = 0.25n^2 - n + 1$$

In the second case, $\phi_1 = 1/1.5 = 0.67$ and $\phi_2 = 1/2.5 = 0.40$.

$$\frac{k_n}{k_H} = (1 - n + 0.67n)(1 - n + 0.4n) = (1 - 0.33n)(1 - 0.60n) = 0.20n^2 - 0.93n + 1$$

Plotting the two quadratics gives:

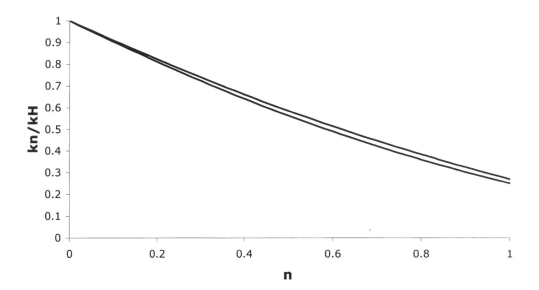

These curves are clearly very similar, with the rate ratios differing by only 7% at the most. Very accurate rate measurements would be needed to distinguish these situations.

27. Though this value appears small, it represents a large k_{12}/k_{14} KIE. This indicates that the C–C bond is cleaved in the rate-determining step of the Baeyer-Villager oxidation.

28. The reaction shown is a 1,2-phenyl shift. The general strategy is to incorporate a reactive group such that two intramolecular rearrangements of the free radical are possible: the phenyl shift and a rearrangement for which the rate constant is already known – the free radical clock. The ratio of observed products allows easy computation of the unknown rate constant. In reality, the rate constant for the clock reaction in the particular species is unknown, but can be presumed to be close to that of the parent system.

As an example, the cyclization of 5-hexenyl radical (Eq. 8.73) can be used as the clock reaction:

Ph shift   cyclization

The product mixture might contain more than two products, but each product should be identified, if possible, as coming from either the phenyl shift or cyclization. The phenyl shift rate constant can then be computed:

$$\frac{k_{\text{Ph shift}}}{k_{\text{cyclization}}} = \frac{\%\ \text{products from Ph shift}}{\%\ \text{products from cyclization}}$$

$$k_{\text{Ph shift}} = k_{\text{cyclization}}\frac{\%\ \text{products from Ph shift}}{\%\ \text{products from cyclization}} = 1.0 \times 10^5\,\text{s}^{-1}\,\frac{\%\ \text{products from Ph shift}}{\%\ \text{products from cyclization}}$$

29, We can paraphrase the first statement of the exercise as follows: The sensitivity of an S_N2 reaction rate to the basicity of the nucleophile is greater if the leaving group is poor. This statement makes intuitive sense – with a poor LG, a good Nuc would likely be needed to make the reaction go, but with a good LG, it shouldn't matter as much whether the Nuc were good or poor.

A bit deeper analysis can be based on the Hammond Postulate (Section 7.3.1). As the LG gets poorer, the energy of the products increases, and a later transition state is expected. This increases the extent of electron donation from the Nuc at the transition state (exaggerated in the illustration below). Thus, the ability of the Nuc to donate electrons (the Nuc's basicity) becomes more important.

$$Nuc:^{\ominus} \quad + \quad \text{C}-LG \longrightarrow \left[^{\delta\ominus}Nuc\text{-----C-----}LG^{\delta\ominus} \right]^{\ddagger} \quad vs. \quad \left[^{\delta\ominus}Nuc\text{-----C-----}LG^{\delta\ominus} \right]^{\ddagger}$$

good LG   poor LG

Similarly paraphrasing the second statement: The sensitivity of the S_N2 rate to the leaving ability of the LG is greater if the Nuc is weaker. This also makes intuitive sense, and a completely analogous Hammond argument may be used here. As the Nuc gets weaker, the energy of the reactants decreases, and a later transition state is expected. The electron donation from the C–LG bond to LG is increased, so the leaving ability of LG is more important.

GOING DEEPER

Our analysis of these statements could go much deeper than the simple Hammond Postulate considerations presented above. Marcus theory (Section 7.7), relating reaction rates to the shapes of energy surfaces, offers additional insights (that we will not attempt to explore here). Qualitative and quantitative analyses of the types of interdependencies addressed in this exercise, and not just for S_N2 reactions, have been presented by many chemists. The name Bema Hapothle was coined to give credit to at least several of the chemists who contributed to understanding of such effects based on transition state structure. This acronym honors Bell, Marcus, Hammond, Polanyi, Thornton, and Leffler for their contributions that spanned more than three decades, starting in the 1930s. See W. P. Jencks, *Chem. Rev.* **1985**, *85*, 511.

9

Catalysis

1. Let us rewrite the chemical equation with the following abbreviations:

$$R + H_2O \underset{k_{-1}[HB^+]}{\overset{k_1[B]}{\rightleftharpoons}} I \xrightarrow{k_2[HB^+]} P + HOMe$$

If we presume that the second step is the rate-determining step, the rate can be written as

$$\frac{d[P]}{dt} = k_2[I][HB^+]$$

We can generate an expression for [I] by using the steady-state approximation:

$$\frac{d[I]}{dt} = 0$$

$$k_1[R][B] - k_{-1}[I][HB^+] - k_2[I][HB^+] = 0$$

$$[I] = \frac{k_1[R][B]}{(k_{-1} + k_2)[HB^+]}$$

Plugging this into the rate law gives

$$\frac{d[P]}{dt} = \frac{k_1 k_2[R][B]}{k_{-1} + k_2} = k_{obs}[R][B]$$

In this substitution, [HB$^+$] drops out of the rate law, and we obtain a rate law with the same concentration dependence as the one we would get by assuming that the first step is rate-determining. Therefore, this overall reaction appears to be catalyzed only by the general base.

We can understand this result by noting that the first step, and acid-base equilibrium occurring before the rate-determining step, cannot really be general base catalyzed but must instead be specific base catalyzed. So the reaction is really catalyzed by specific base and general acid, and this situation is kinetically equivalent to general base catalysis (see Section 9.3.3).

2. Starting from Eq. 9.23,

$$\frac{d[P]}{dt} = k[R^-]$$

We can obtain an expression for [R$^-$] from the acid-base equilibrium:

$$K_{aHR} = \frac{[R^-][H_3O^+]}{[RH]}$$

$$[R^-] = \frac{K_{aHR}[RH]}{[H_3O^+]}$$

Plugging this in, we obtain Eq. 9.24:

$$\frac{d[P]}{dt} = \frac{kK_{aHR}[RH]}{[H_3O^+]}$$

3. The rate law for the reaction shown is

rate $= k[\text{acetone}][B]$

We can calculate k by using Eq. 9.37:

$$\log(k) = \beta \bullet pK_a + C'$$

Since we want relative rates, we will set up our calculation that way:

$$\frac{\text{rate}_2}{\text{rate}_1} = \frac{k_2[\text{acetone}][B]_2}{k_1[\text{acetone}][B]_1} = \frac{k_2[B]_2}{k_1[B]_1}$$

Notice that the relative rates will depend not only on the relative rate constants, but also the relative concentrations of the base present. We can get the relative k values by manipulation of Eq. 9.37. (Recall that the k in Eq. 9.37 is the same k from the rate law, not a k_{obs} that incorporates concentrations.)

$$\log(k_2) - \log(k_1) = (\beta_2 \bullet pK_{a2} + C') - (\beta_1 \bullet pK_{a1} + C')$$

$$\log(\frac{k_2}{k_1}) = \beta_2 \bullet pK_{a2} - \beta_1 \bullet pK_{a1}$$

$$\frac{k_2}{k_1} = e^{\beta_2 \bullet pK_{a2}} e^{-\beta_1 \bullet pK_{a1}}$$

We can get the [B] values from the Henderson-Hasselbalch (H-H) equation:

$$pH = pK_a + \log\left(\frac{[B]}{[BH^+]}\right)$$

For our relative rate calculation, we do not need (nor do we have) any absolute concentrations. Rather, the fraction of [B] and [BH$^+$] present as [B] will suffice:

$$\text{Fraction B} = \frac{[B]}{[B]+[BH^+]} = \frac{1}{([B]+[BH^+])/[B]} = \frac{1}{1+([BH^+]/[B])}$$

We could then get the acid/base ratio from the H-H equation and then plug it in here. Since we have three different pK_a values, let rearrange the H-H equation and plug it in:

$$\frac{[B]}{[BH^+]} = 10^{pH-pK_a}$$

$$\frac{[BH^+]}{[B]} = 10^{pK_a-pH}$$

$$\text{Fraction B} = \frac{1}{1+\left(10^{pK_a-pH}\right)} = \frac{1}{1+\left(10^{pK_a-7}\right)}$$

Now we're all set. We'll calculate the relative rate constants (taking always k_1 as the constant for the reaction with pH = 4 and β = 0.2), the fraction of B present for each pK_a value, and the relative rates:

	k_2/k_1			Fraction B	Rel. rate		
	$\beta = 0.2$	$\beta = 0.5$	$\beta = 0.9$		$\beta = 0.2$	$\beta = 0.5$	$\beta = 0.9$
$pK_a = 4$	1.0	3.3	16.4	0.999	1.0	3.3	16.4
$pK_a = 7$	1.8	14.9	244.7	0.500	0.9	7.4	122.5
$pK_a = 10$	3.3	66.7	3641.0	0.001	0.003	0.1	3.6

The fastest rate is obtained for the medium base and the highest β value. It's clear that the fastest rate is *not* obtained with the strongest base for any of the β values in this example with pH = 7. The stronger the base, the less of it that is present in solution at a given pH. If the pK_a is significantly greater than the pH, the amount of base in solution is very small.

GOING DEEPER

It may seem very counterintuitive that making a base stronger in a base-catalyzed reaction could actually slow down the reaction. The problem here might be a lack of appreciation of one constraint of the exercise: that the pH stays the same as we change pK_a. Let's think about what this means experimentally. Suppose we prepare 1.0 M solutions of two bases that have conjugate acids with $pK_a = 4$ and 10. The pH of the first solution would be 9, and the pH of the second would be 12. (If you are unsure how to calculate this, consult any general chemistry textbook.) So if we wanted to do the acetone hydration study as described, what would we need to do? We would have to add acid to the solutions in order to adjust the pH to 7. But more acid would be required for the solution of the stronger base – in fact, 1,000 times more. For each liter of solution, approximately 1 mmol of HCl would be required for the weaker base, while approximately 1 mol would be needed for the stronger base. Realizing this, the conclusion of this exercise no longer seems so strange. Using a stronger base requires us to add more acid – that's why the base-catalyzed reaction can be slower!

4. The rate plateau in the pH range from 4-10 and the decline in rate from pH 4 to pH 2 are characteristic of general base catalysis with the conjugate base of an acid with pK_a of approximately 4 (see Figure 9.9). The increase in rate above pH 10 can be attributed to specific base catalysis, and the increase in rate below pH 2 can be attributed to specific acid catalysis.

If we call the acid with pK_a near 4 HA, then a rate law consistent with the overall behavior would be:

$$\frac{d[P]}{dt} = k_1[R][H_3O^+] + k_2[R][A^-] + k_3[R][OH^-]$$

5. The Hammett equation is

$$\ln\left(\frac{k}{k_0}\right) = \rho\sigma$$

Eqs. 9.36 and 9.37 differ from this format in that they are internally referenced – not referenced to a different reaction. There is no reason, however, that we cannot do this. Let's suppose that we've picked a reference reaction. Eq. 9.36 for this reaction would be

$$\log(k_0) = -\alpha_0 \bullet pK_a + C$$

Note that the sensitivity (α) is expected to differ for the reference reaction, just as in the Hammett equation. However, here the value of α_0 is already defined; we do not have the freedom to set its value to 1. Subtracting this equation from Eq. 9.36 gives

$$\log(k) - \log(k_0) = \left(-\alpha \bullet pK_a + C\right) - \left(-\alpha_0 \bullet pK_a + C\right)$$

$$\log\left(\frac{k}{k_0}\right) = \left(\alpha_0 - \alpha\right)pK_a$$

This equation is in essentially the same format as the Hammett equation, except that the sensitivity of the reference reaction shows up here. Doing the same thing for Eq. 9.37:

$$\log(k) - \log(k_0) = \left(\beta \bullet pK_a + C'\right) - \left(\beta_0 \bullet pK_a + C'\right)$$

$$\log\left(\frac{k}{k_0}\right) = \left(\beta - \beta_0\right)pK_a$$

6. Acid catalysis of the tautomerization would occur by protonation on N, followed by deprotonation at C. Specific catalysis would require that the imine be fully protonated in a fast equilibrium, followed by a slow deprotonation. Significant protonation of the imine would require that the pH be comparable to the imine pK_a (4-5), and this pH could be achieved only with an acid having a comparable or lower pK_a. The pK_a values of HCl, picric acid, acetic acid, ammonium, and phenol are –6.1, 0.4, 4.76, 9.24, and 10.02, respectively (Tables 5.5 and 5.7). Therefore, specific acid catalysis could potentially be caused by HCl, picric acid, or acetic acid.

General catalysis would require direct protonation of the imine by the acid catalyst in the rate determining step. According to the Libido rule, the protonation should be more favorable in the transition state than in the ground state. This would not be possible with either HCl or picric acid, because these would fully protonate the imine reactant. Possible general acids are acetic acid, ammonium, and phenol, all of which have pK_a values comparable to or higher than that of the imine.

7. Several strategies are available for catalysis:
   a. Place charges that will electrostatically stabilize the transition state (negative charges near the positive N, positive charges near the negative O).
   b. Place a general acid near the OTs within H-bonding distance, so that the negative charge in the transition state will be further stabilized.
   c. Create a cavity that will accommodate the reactant but that will provide an even better fit for the transition state. More specifically, some steric strain can be applied that will distort the methyl groups toward their positions in the transition state.

   d. Place the cavity for the OTs group slightly farther than optimum for the reactant, encouraging the N–O bond to stretch.

8. In the mechanism below, note that several parts of the pyridoxal phosphate molecule can be involved in the reaction, aiding the catalysis: the aldehyde (forming imines), the hydroxy group (increasing electrophilicity of the aldehyde, stabilizing and rigidifying imine intermediates), and the positive nitrogen (increasing electrophilicity of the aldehyde, promoting proton transfers).

9. The HOBT serves as a nucleophilic catalyst, forming an activated ester. Hydroxylamines (as well as hydrazines and other compounds with adjacent heteroatoms) are very active nucleophiles. Also, the leaving group ability of ⁻OBT is much greater than that of an alkoxide, being much less basic (similar to a phenoxide).

10. a. Specific acid:

b. General acid:

c. Specific base:

d. General base:

11. The reverse of the general-acid catalyzed reaction in 10b would also be general-acid catalyzed if the first step were rate-determining. However, the principle of microscopic reversibility says that the same path is traversed in both directions, so the highest energy point should be the same in both directions. Therefore, the second step of the reverse reaction is expected to be rate determining. In this case, the reaction should be specific-acid and general-base catalyzed, a kinetically equivalent situation to general-acid catalysis. (Recall from Section 9.3.1 that an equilibrium before the rate-determining step that involves a general acid or base is kinetically equivalent to specific-acid or base catalysis. Also, recall from Section 9.3.3 the kinetic equivalence of general-acid catalysis and specific-acid/general-base catalysis.)

12. Incorporating the acid and base concentrations into the rate constant gives the same rate law for each term: rate = k_{obs}[acetone]. Then the pH dependence comes from the effect of pH on the concentration of each acid or base:

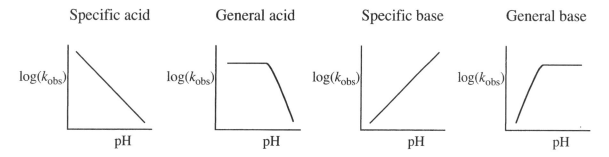

13. Looking back at our general-acid and general-base catalyzed mechanisms, it is clear that they each involve the same two steps, differing only in the order of the steps. The most straightforward mechanism that involves both a general acid and a general base would be one in which both steps occur simultaneously. Even in this case, a second step is still necessary to regenerate the catalysts:

Examining each of the experimental results, we find that all are consistent with this mechanism. First, the rate dependence on acetone, HA, and B is consistent.

i. The solvent isotope effect can be understood by recognizing that HA would exchange with D_2O to form DA. So, even though solvent does not appear in our mechanism as drawn, we can interpret the solvent isotope effect as showing that $k_{HA}/k_{DA} = 2.0$. This is consistent with a primary isotope effect coming from cleavage of the H–A or D–A bond in the rate-determining step.

ii. The acetone isotope effect is also primary, as expected since the C–H bond is cleaved in the rate-determining step: $k_{H-acetonyl}/k_{D-acetonyl} = 5.8$. (You might be tempted to adjust this value to correct for the fact that all six positions are deuterated in order to obtain a per-bond isotope effect. However, the reaction involves cleavage of only one C–H bond, so no such correction is necessary. The other deuteria will lead to small secondary effects that are included in the value of 5.8, but these are small relative to the primary effect.)

The fact that this isotope effect is larger than the other primary effect suggests that this proton is approximately halfway transferred in the activated complex, while the transition state is either early or late relative to the protonation by HA.

iii. The α value of 0.2 and β value of 0.88 help to clarify the comment just made. The low α value shows that the transition state is indeed early with respect to the protonation by HA. That is, at the transition state, the proton is still mostly bonded to A rather than the oxygen. However, the high value of β shows that the action of B is much more significant at the transition state and even that the deprotonation is mostly complete. This result is a bit at odds with the large isotope effect. One might have expected a smaller isotope effect if the transition state is really this late with respect to the deprotonation (see Figure 8.5).

14. a.

b.

15. i. The stereospecificity is consistent only with the 14b mechanism. In this mechanism, the anomeric C–O bond is never broken, preserving the starting stereochemistry. In the mechanism of part a, stereochemistry of the product is determined by the attack of water on the oxocarbenium ion. While complete stereoselectivity for the α or β product is possible in this reaction, the common intermediate (the oxocarbenium ion) from the α or β reactant requires that the same product (or mixture of products) be formed in each case. Thus, this mechanism is eliminated by the stereospecificity of the reaction.

ii. This isotope labeling is consistent only with the mechanism in 14b. Since no C–O bonds of the glucose product are cleaved in the mechanism, there is no opportunity for ^{18}O incorporation. However, the carbonyl oxygen atom in acetone comes from water, so ^{18}O incorporation occurs. In the part a mechanism, one hydroxyl of glucose comes from water, so incorporation of the label into glucose would be expected from this mechanism. The incorporation of label into acetone, though not anticipated through this mechanism, would be expected anyway through hydration-dehydration, which occurs rapidly in aqueous acid. (In other words, the labeling result concerning acetone offers no mechanistic information.)

16. First, it is important to note that the formation of the acetal from methanol and the deuterium labeling results are fully consistent with the 14b mechanism. The four results given further show that the first step shown in 14b is irreversible and rate-determining. Being an acid-base reaction, the first step would be expected to be reversible unless it were rate-determining.

i and ii. If the first step were reversible, some product molecules (either the acetal or acetone) would be expected to have more than one deuterium. That is, reaction with D^+ could be followed by loss of H^+ and reaction again with D^+:

iii. The primary isotope effect also shows that the rate-determining step involves cleavage of a bond to H or D, as happens in the first step.

iv. The high α value shows that protonation by H_3O^+ is involved in the rate-determining step, also consistent with a rate-determining first step.

Even though the acid is the specific acid (H_3O^+), the fact that the protonation occurs in the rate-determining step means that H_3O^+ actually acts as a general acid in this reaction.

17. Increasing the acid strength causes the bottom edge of the More O'Ferrall plot to be lowered in energy. This moves the transition state roughly parallel to the bottom edge, the result of movement off the diagonal toward the more stable intermediate (toward the lower left corner) and movement along the diagonal to make the transition state earlier. Thus, the extent of proton transfer at the transition state stays approximately the same.

18. All three scenarios can be solved in the same way that the non-inhibited reaction was solved (Eqs. 9.49-52). In all three, the rate-determining step is still the k_{cat} step, so Eq. 9.49 still holds:

$$\frac{d[P]}{dt} = k_{cat}[ES]$$

The strategy will be to consider all forms of the enzyme and then solve for [E], use the steady state approximation on ES to solve for [ES], and then substitute the result into the above rate law.

a. **Competitive inhibition**

$$[E]_0 = [E] + [ES] + [EI]$$

We can use the inhibition equilibrium to solve for [EI]:

$$K_{\mathrm{I}} = \frac{[\mathrm{EI}]}{[\mathrm{E}][\mathrm{I}]}$$

$$[\mathrm{EI}] = K_{\mathrm{I}}[\mathrm{E}][\mathrm{I}]$$

$$[\mathrm{E}]_0 - [\mathrm{ES}] = [\mathrm{E}] + K_{\mathrm{I}}[\mathrm{E}][\mathrm{I}]$$

$$[\mathrm{E}] = \frac{[\mathrm{E}]_0 - [\mathrm{ES}]}{K_{\mathrm{I}}[\mathrm{I}] + 1}$$

The steady-state approximation for ES gives:

$$\frac{d[\mathrm{ES}]}{dt} = k_1[\mathrm{E}][\mathrm{S}] - k_{-1}[\mathrm{ES}] - k_{\mathrm{cat}}[\mathrm{ES}] = 0$$

$$k_1\left(\frac{[\mathrm{E}]_0 - [\mathrm{ES}]}{K_{\mathrm{I}}[\mathrm{I}] + 1}\right)[\mathrm{S}] - k_{-1}[\mathrm{ES}] - k_{\mathrm{cat}}[\mathrm{ES}] = 0$$

$$\frac{k_1[\mathrm{E}]_0[\mathrm{S}]}{K_{\mathrm{I}}[\mathrm{I}] + 1} = \frac{k_1[\mathrm{ES}][\mathrm{S}]}{K_{\mathrm{I}}[\mathrm{I}] + 1} + k_{-1}[\mathrm{ES}] + k_{\mathrm{cat}}[\mathrm{ES}]$$

Before going further, this is a good time to simplify by dividing both sides by k_1, multiplying both sides by $(K_{\mathrm{I}}[\mathrm{I}]+1)$, and introducing the Michaelis constant, $K_{\mathrm{M}} = (k_{\mathrm{cat}} + k_{-1})/k_1$.

$$[\mathrm{E}]_0[\mathrm{S}] = [\mathrm{ES}][\mathrm{S}] + K_{\mathrm{M}}[\mathrm{ES}]\big(K_{\mathrm{I}}[\mathrm{I}] + 1\big)$$

$$[\mathrm{ES}] = \frac{[\mathrm{E}]_0[\mathrm{S}]}{[\mathrm{S}] + K_{\mathrm{M}}\big(K_{\mathrm{I}}[\mathrm{I}] + 1\big)}$$

$$\frac{d[\mathrm{P}]}{dt} = \frac{k_{\mathrm{cat}}[\mathrm{E}]_0[\mathrm{S}]}{[\mathrm{S}] + K_{\mathrm{M}}\big(K_{\mathrm{I}}[\mathrm{I}] + 1\big)}$$

Taking $V_{\mathrm{max}} = k_{\mathrm{cat}}[\mathrm{E}]_0$, we get

$$\frac{d[\mathrm{P}]}{dt} = \frac{V_{\mathrm{max}}[\mathrm{S}]}{[\mathrm{S}] + K_{\mathrm{M}}\big(K_{\mathrm{I}}[\mathrm{I}] + 1\big)}$$

This rate law is the same as that without inhibition (Eq. 9.52) except for the $(K_{\mathrm{I}}[\mathrm{I}]+1)$ factor in the denominator. Since $K_{\mathrm{I}}[\mathrm{I}] > 0$, it must be true that $K_{\mathrm{M}}\big(K_{\mathrm{I}}[\mathrm{I}] + 1\big) > K_{\mathrm{M}}$. The $(K_{\mathrm{I}}[\mathrm{I}]+1)$ factor thus decreases the rate and requires a larger [S] to achieve the saturation condition, in which $[\mathrm{S}] \gg K_{\mathrm{M}}\big(K_{\mathrm{I}}[\mathrm{I}] + 1\big)$ and the rate equals V_{max}. Thus, the inhibitor competes with S, slows the reaction, and makes saturation less likely. With enough S, however, S can effectively compete with I, such that a rate of V_{max} is possible.

b. **Noncompetitive inhibition**

$$[E]_0 = [E] + [ES] + [EI] + [ESI]$$

We use the two inhibition equilibria to solve for [EI] and [ESI]:

$$K_I = \frac{[EI]}{[E][I]} \text{ and } K_I = \frac{[ESI]}{[ES][I]}$$

$$[EI] = K_I[E][I] \text{ and } [ESI] = K_I[ES][I]$$

$$[E]_0 - [ES] - K_I[ES][I] = [E] + K_I[E][I]$$

$$[E] = \frac{[E]_0 - [ES] - K_I[ES][I]}{K_I[I] + 1}$$

In this case, the steady state approximation for ES should have two additional terms that represent the equilibrium with ESI. For simplicity, we will make the additional approximation that these terms cancel each other. In effect, we are assuming that this equilibrium is established throughout the reaction. (Remember that the forward and reverse rates of an equilibrium are equal.) Plugging [E] into the steady-state equation from part a gives:

$$k_1 \left(\frac{[E]_0 - [ES] - K_I[ES][I]}{K_I[I] + 1} \right)[S] - k_{-1}[ES] - k_{cat}[ES] = 0$$

$$\frac{k_1[E]_0[S]}{K_I[I] + 1} = \frac{k_1[ES][S](1 + K_I[I])}{K_I[I] + 1} + k_{-1}[ES] + k_{cat}[ES]$$

$$\frac{k_1[E]_0[S]}{K_I[I] + 1} = k_1[ES][S] + k_{-1}[ES] + k_{cat}[ES]$$

As in part a, we now simplify by dividing both sides by k_1 and introducing the Michaelis constant.

$$\frac{[E]_0[S]}{K_I[I] + 1} = [ES][S] + K_M[ES]$$

$$[ES] = \frac{[E]_0[S]}{([S] + K_M)(K_I[I] + 1)}$$

$$\frac{d[P]}{dt} = \frac{V_{max}[S]}{([S] + K_M)(K_I[I] + 1)}$$

This rate law resembles the one in part a, but here the $(K_I[I]+1)$ factor affects the whole denominator. Again the presence of this factor decreases the rate, but now saturation and a rate of V_{max} is not possible if $[I] > 0$.

c. **Nonproductive inhibition**

$$[E]_0 = [E] + [ES] + [ES']$$

$$K_{eq} = \frac{[ES']}{[E][S]}$$

$$[ES'] = K_{eq}[E][S]$$

$$[E]_0 - [ES] = [E] + K_{eq}[E][S]$$

$$[E] = \frac{[E]_0 - [ES]}{K_{eq}[S] + 1}$$

Plugging into the steady-state approximation for ES gives:

$$k_1\left(\frac{[E]_0 - [ES]}{K_{eq}[S] + 1}\right)[S] - k_{-1}[ES] - k_{cat}[ES] = 0$$

$$\frac{k_1[E]_0[S]}{K_{eq}[S] + 1} = \frac{k_1[ES][S]}{K_{eq}[S] + 1} + k_{-1}[ES] + k_{cat}[ES]$$

We now simplify by dividing both sides by k_1, multiplying both sides by $(K_{eq}[S]+1)$, and introducing the Michaelis constant.

$$[E]_0[S] = [ES][S] + K_M[ES]\left(K_{eq}[S] + 1\right)$$

$$[ES] = \frac{[E]_0[S]}{[S] + K_M\left(K_{eq}[S] + 1\right)}$$

$$\frac{d[P]}{dt} = \frac{V_{max}[S]}{[S] + K_M\left(K_{eq}[S] + 1\right)}$$

This rate law (and the whole calculation) is the same as in part a except for the replacement of $K_I[I]$ with $K_{eq}[S]$. If K_{eq} is significant, the rate will be decreased relative to the non-inhibited reaction. In this case, additional insight can be gained by rearranging the denominator:

$$\frac{d[P]}{dt} = \frac{V_{max}[S]}{[S] + K_M K_{eq}[S] + K_M} = \frac{V_{max}[S]}{[S]\left(1 + K_M K_{eq}\right) + K_M}$$

Now we can see that if $[S]\left(1 + K_M K_{eq}\right) \gg K_M$, the rate will then equal $V_{max}\big/\left(1 + K_M K_{eq}\right)$. Unless $K_M K_{eq} \ll 1$, a rate of V_{max} will not be obtained.

19. In going from the first compound to the second, rotation about the α–β bond relative to the CO₂H group is eliminated, serving to hold the CO₂H and OH groups in closer proximity. In the third compound, three methyl groups replace three H's. Through non-bonded repulsions, the methyl groups should push the CO₂H and OH groups incrementally closer. Rotation between the alkenyl and aryl groups is also more hindered, and the average dihedral angle might be changed to increase the proximity of the CO₂H and OH groups.

2.3 x 10⁷   3.7 x 10¹¹   2 x 10¹³

10

Organic Reaction Mechanisms Part 1: Reactions Involving Additions and/or Eliminations

S O L U T I O N S   T O   E X E R C I S E S

1.	Nucleophile	Electrophile	Lewis basic site	Lewis acidic site	HOMO or LUMO
A	O lone pairs		O		HOMO
B		C attached to Cl			LUMO
C	O lone pairs	H attached to O	O	H	HOMO LUMO

	Nucleophile	Electrophile	Lewis basic site	Lewis acidic site	HOMO or LUMO
D	O lone pairs	C attached to O	O	C	HOMO LUMO
E	π bond		π bond		HOMO
F	N lone pair	C attached to N	N	C	HOMO LUMO
G		C attached to O, H attached to O		C, H	LUMO LUMO+1

The meaning of nucleophile and Lewis basic site is very similar, the difference being that the former is a kinetic term and the latter is generally a thermodynamic term. For all of these molecules, the sites that can donate electrons during a reaction can also give complexes or products through donation of the electrons to an appropriate Lewis acid. Likewise, most of the electrophiles, able to accept electrons during a reaction, are also able to form complexes with appropriate Lewis bases. The one exception is ethyl chloride, which does not complex with Lewis bases without loss of the chloride. (The complex would be the S_N2 transition state, which is not stable.) Acidic hydrogens can complex with Lewis bases in hydrogen bonding. The HOMO and LUMO diagrams show the most available concentrations of electrons and electron deficiency and serve as predictors of both kinetic and thermodynamic reactivity. Note that only reasonably good nucleophiles and electrophiles are listed above. For example, the lone pairs on Cl in b and on O in g are very weakly nucleophilic.

2. One clear change in the overall reaction is the location of the thiol H. Since thiols are more acidic than alcohols, we might expect this H to experience greater bonding in the alcohol. This idea is consistent with the observation of higher equilibrium constants.

Analysis of BDEs can support this idea. The average BDE for the S–H bond is 90 kcal/mol, which is 14 less than the BDE for the O–H bond in CH_3OH of 104.4 (Table 2.2). This will tend to make the equilibrium constant higher in the thiol case. The BDEs for C–S and C–O also differ, but this comparison appears to predict the opposite effect. The average BDE of 60 for the C–S bond is significantly lower than the C–O BDE of 83.2 for CH_3OCH_3, and since these bonds are on the product sides, this would tend to disfavor thiol addition. A possible explanation is that the BDEs cited might not well represent the bonds in these acetal structures. While the C–S and C–O BDE difference would likely counteract the difference between S–H and O–H, a $\Delta\Delta G$ of only 4 kcal/mol is all that is needed to explain an equilibrium constant ratio of 10^3.

3. Thiols are much more acidic than alcohols, such that significant amounts of the thiolate (RS⁻) can be present at pH values near the neutral range. Since thiolates are much more nucleophilic than thiols, the addition reaction can be dominated by the thiolate, even if it is the minor conjugate at a given pH. Addition of a thiolate to a ketone is assisted by a general acid, since without it a highly basic alkoxide would be required as an intermediate. The equilibrium constant for the acid-free addition step would therefore strongly favor the reactants. The general acid could be the thiol itself.

4. These substitution reactions occur stereospecifically to give the retention product, often a sign of neighboring group assistance. (Both reactions are shown with wedged and dashed bonds. For clarity, the second reaction is shown also with Newman projections.)

5. The observation of trans dibromides suggests that both products come from bromonium ion intermediates. Since the two faces of the alkene are diastereotopic, two bromonium ions are possible. Further, in each bromonium ion, the two sites for nucleophilic attack are different. The steric bulk of the *t*Bu group apparently affects the selectivity of one or both of these steps. Since the *t*Bu will clearly prefer the equatorial position, its relatively remote effect on the selectivities is not immediately obvious. Given this, it seems most convenient to first catalog the products expected from all possible paths.

The trans bromonium ion (with the Br and *t*Bu on opposite faces of the ring) can give either the diaxial (major) or diequatorial (minor) products, depending on whether the tribromide (or bromide) attacks at C1 or C2. (For clarity, the products are shown here in the same conformation as the starting cyclohexene. Upon relaxation into the chair form with equatorial *t*Bu, the bromines in the first product would both be axial while those in the second product would both be equatorial.)

The cis bromonium ion can give the same two products, though nucleophilic attack is required at the opposite C from the trans case in order to get the same product.

So far, we have shown two paths to each product. A simplistic explanation for the product ratio is just that the selectivity in both steps is determined by the incoming group's preference to stay as far from the *t*Bu group as possible. The trans bromonium ion would then be favored, as would attack on C1, and this indeed would give the major product. The minor product could then come from imperfect selectivity in either step.

In order to make a more carefully considered argument, constructing models is a good idea. From this exercise, one will likely find that the expected selectivities are still not obvious, but that some conclusions can be drawn. One effect of the bulky *t*Bu is to push the geminal H (the one on C4), already axial, further toward the center of the top face of the ring. The corresponding axial H on the bottom face at C5 is not pushed towards the center. Thus, the top face might be expected to be somewhat protected from attack, favoring formation of the cis bromonium ion. But nucleophilic attack on the cis isomer must then be from the top face, and the preferred site for attack would seem to be C1, being further from the axial H on C4. This would give the observed minor product. A literature suggestion is that the formation of the bromonium ions is reversible and that even if the cis isomer is favored, nucleophilic attack on this isomer is slower, such that the nucleophilic step occurs primarily from the trans bromonium ion and primarily at C1, giving the diaxial dibromide. (Barili, P. L.; Bellucci, G.; Marioni, F.; Morelli, I.; Scartoni, V. *J. Org. Chem.* **1972**, *37*, 4353.)

6. The addition goes with *anti* stereochemistry, going through the bromonium ions, such that the products shown are obtained:

7. The quadrupole moment of aromatic rings allows for significant stabilization of positive charges near the face of the rings through cation-π interactions. Some other amino acids also have electron-donating groups such as OH, NH_2, or SH groups, but these groups would more likely intercept the carbocation irreversibly. Thus, the aromatic groups can both stabilize the cation intermediates and shield them from the environment.

8. The following mechanism is probably the simplest:

The expected rate law for this pathway, taking the hydride transfer step to be rate limiting, is

Rate = $K_{eq}k[RCHO]^2[OH^-]$

Aside from experimentally determining the rate law, other possible mechanistic tests include (1) isotope labeling: use RCDO, which should give RCD_2OH by the above mechanism; (2) kinetic isotope effect: use RCDO and measure k_H/k_D, which should be a primary effect for the above mechanism; and (3) LFER: vary X in p-XC_6H_4CHO and plot log k vs. σ, which should give a small ρ value (since a charge is transferred from an O β to R in one molecule to an O β to R in another molecule).

An alternative mechanism is

The rate law for this mechanism is indistinguishable from the other one:

Rate = $K_{eq}K_{eq}'k'[RCHO]^2[OH^-]$

However, if this mechanism were correct, then the results of most of our other tests would be different: (1) formation of RCHDOH; (2) primary KIE; and (3) large, positive ρ value.

9.

The rate-determining step is most likely attack of the weakly nucleophilic water on the iminium ion. The identity of "B" in this mechanism could be water, imine, hydroxylamine, etc.

10. Three factors might contribute. Placing the negative charge next to the Cl atoms allows for a hyperconjugative interaction that delocalizes the charge to F. (The conformation shown results after rotation about the C–C bond. The conformation formed without rotation has weaker hyperconjugative interactions but involving both C–F bonds.) Second, attack of the nucleophile at the CF_2 end is favored in terms of sterics. Finally, the Cl lone pairs can better avoid an adjacent C lone pair than can the F lone pairs, since the C–Cl bonds are longer and Cl is more polarizable. (This last effect was used to explain the higher acidity of $HCCl_3$ relative to HCF_3 in Section 5.3.2.)

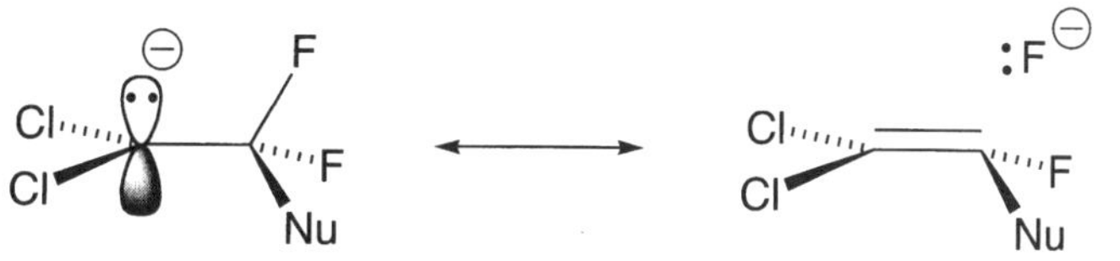

11. The mechanism is a radical chain. Initiation is through photochemical bond cleavage:

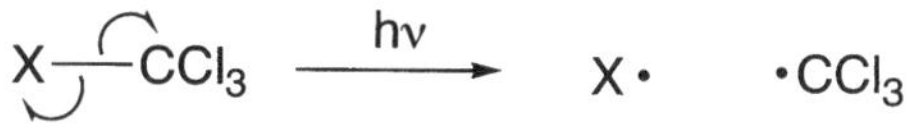

Propagation steps:

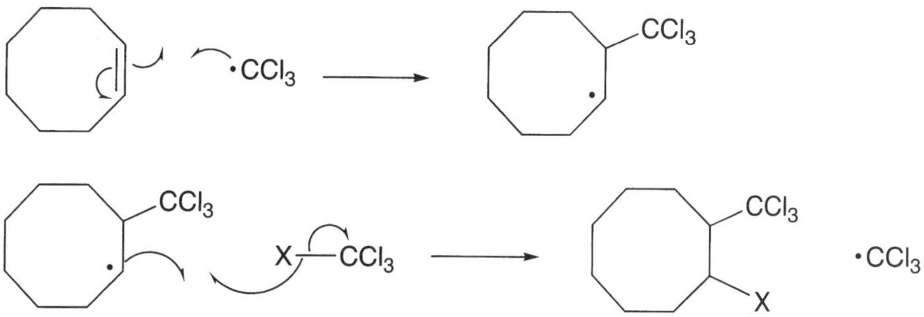

The following intramolecular propagation step can also compete:

This leads to the 1,4-product through one more propagation step analogous to the second one above. The ability of the 1,5-H abstraction to compete with halide abstraction depends on the rate of halide abstraction. Br is more readily abstracted, allowing less time for the intramolecular rearrangement. Various termination steps are possible.

12. The regiochemistry suggests the intermediacy of a carbocation that is resonance-stabilized by the Ph group (as in Eq. 10.23); protonation at the other carbon would give a non-stabilized cation. The *syn* stereochemistry is suggestive of the contact ion pair explanation of Section 10.4.2 for *syn* addition of HCl to alkenes: the ion pair collapses before the cation has enough time to rotate and allow attack of chloride from the opposite face. Here, a different explanation might also (or instead) apply: sterics. The attack of chloride next to H gives the *syn* product, while *anti* addition would require attack next to Me.

13. The first stage is the formation of a tosylhydrazone (by the usual imine formation mechanism):

The strong base then deprotonates at N, leading to elimination of Ts⁻ and then N_2 from the resulting 1,1-diazene. The carbene then inserts into the neighboring C–H bond.

1,1-diazene

Tests can be devised for both the 1,1-diazene and carbene intermediates. By running the reaction at low temperature, the unimolecular reaction of the 1,1-diazene might be slowed sufficiently to allow spectroscopic observation. Both IR and visible spectroscopies are suitable. (Even your eyes are suitable, as the 1,1-diazene is purple!) The carbene can be trapped by adding a sufficient concentration of a suitable trapping agent, such that a bimolecular trapping reaction can compete with the unimolecular reaction. Possible trapping agents are alcohols (insertion into the O–H bond) and alkenes (formation of cyclopropanes).

14. Each of the three products results from an intramolecular carbene insertion into a C–H bond. The carbene and the three C–H bonds are shown:

15. For convenience, the following abbreviations will be used:

$$\text{ArOEt} + \text{HNuc} \underset{k_{-1}}{\overset{k_1}{\rightleftharpoons} } \text{ArOEt(HNuc)} \underset{k_{-2}}{\overset{B^-, k_2}{\rightleftharpoons}} \text{ArOEt(Nuc)}^- + \text{BH} \overset{k_3}{\longrightarrow} \text{ArNuc} + \text{B}^- + \text{EtOH}$$

Since the last step is rate-limiting,

Rate = $k_3[\text{ArOEt(Nuc)}^-][\text{BH}]$

Taking the steady-state assumption with respect to ArOEt(Nuc)⁻,

$k_2[\text{ArOEt(HNuc)}][\text{B}^-] - k_{-2}[\text{ArOEt(Nuc)}^-][\text{BH}] - k_3[\text{ArOEt(Nuc)}^-][\text{BH}] = 0$

$$[\text{ArOEt(Nuc)}^-] = \frac{k_2[\text{ArOEt(HNuc)}][\text{B}^-]}{(k_{-2} + k_3)[\text{BH}]}$$

Plugging this into the rate law gives

$$\text{Rate} = \frac{k_2 k_3[\text{ArOEt(HNuc)}][\text{B}^-]}{(k_{-2} + k_3)}$$

The initial equilibrium is represented by

$$\frac{k_1}{k_{-1}} = \frac{[\text{ArOEt(HNuc)}]}{[\text{ArOEt}][\text{HNuc}]}$$

Solving for [ArOEt(HNuc)] and substituting into the rate law gives

$$\text{Rate} = \frac{k_1 k_2 k_3[\text{ArOEt}][\text{HNuc}][\text{B}^-]}{k_{-1}(k_{-2} + k_3)} = k_{obs}[\text{ArOEt}][\text{HNuc}][\text{B}^-]$$

A kinetic dependence on HNuc should be observed, since it is involved before the rate-determining step. Through coupled equilibria, it determines the amount of ArOEt(Nuc)⁻ that is available for the last step.

16. The fact that a radical scavenger disrupts the selectivity indicates that the regioselective reaction involves a radical chain process and that in the absence of radicals an elimination-addition mechanism takes over. The radical chain process is likely $S_{RN}1$, initiated by reduction of the aryl iodide:

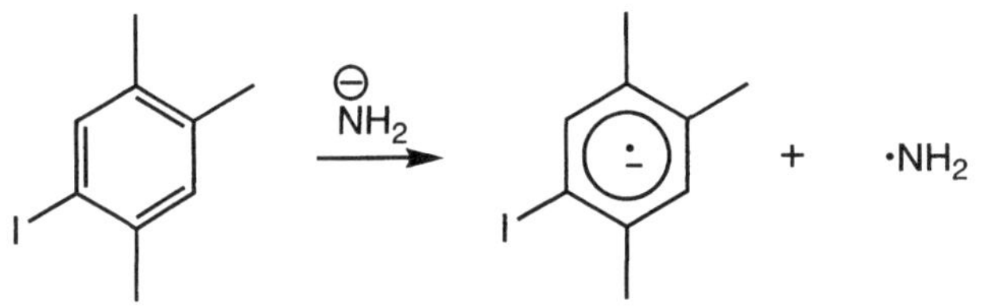

Propagation steps:

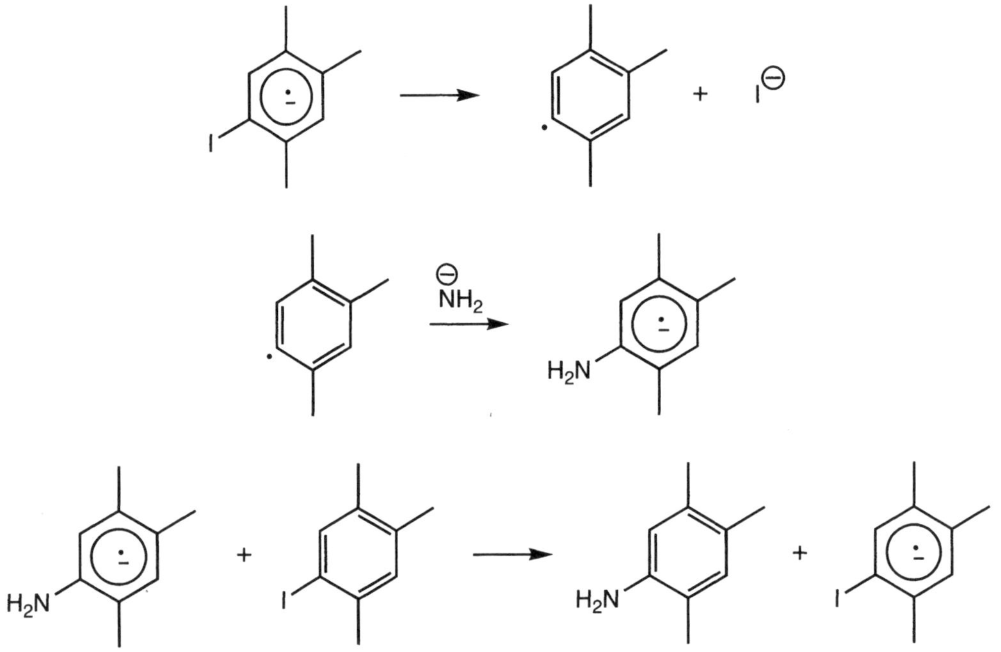

The elimination-addition (benzyne) mechanism is shown below. The nucleophilic attack in the second step can occur at either end of the triple bond.

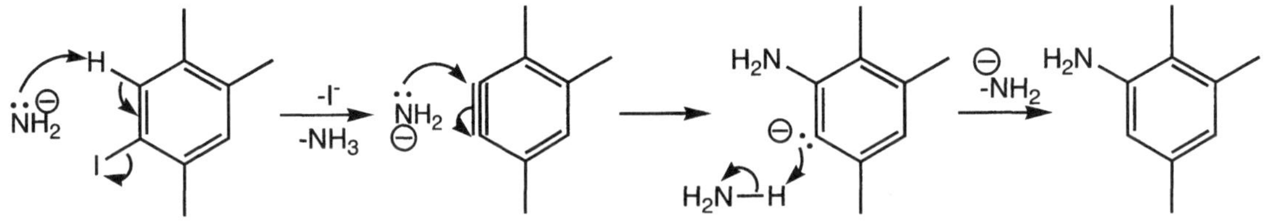

17. An $S_{RN}1$ mechanism is indicated:

Initiation:

$$ArF \quad + \quad {}^{\ominus}OAc \quad \longrightarrow \quad ArF\overset{\bullet}{}{}^{-} \quad + \quad \cdot OAc$$

Propagation:

$$ArF\overset{\bullet}{}{}^{-} \quad \longrightarrow \quad Ar\cdot \quad + \quad F^{\ominus}$$

$$Ar\cdot \quad + \quad {}^{\ominus}OAc \quad \longrightarrow \quad ArOAc\overset{\bullet}{}{}^{-}$$

$$ArOAc\overset{\bullet}{}{}^{-} \quad + \quad ArF \quad \longrightarrow \quad ArOAc \quad + \quad ArF\overset{\bullet}{}{}^{-}$$

18. Section 10.18.2 says that the first step is usually rate-limiting in electrophilic aromatic substitutions.

19. Aluminum chloride is a strong Lewis acid that can abstract halides from alkyl halides to produce carbocations. However, 1° cations are not readily formed, such that the alkylation reaction more likely involves an aluminum chloride complex.

Rearrangement to a more stable 2° cation can compete with the reaction with benzene.

At the higher temperature, the complex more readily rearranges, leading to a higher yield of the 2° product. Unimolecular processes, generally having higher $\Delta S^{\ddagger}$, often compete more favorably with bimolecular processes at higher temperatures. (If two processes have similar rates at one temperature, the one with the higher $\Delta S^{\ddagger}$ will also have a higher $\Delta H^{\ddagger}$, leading to a greater dependence of its rate on temperature. See exercise 19 in Chapter 7.)

20. Before considering stereochemistry, we first note that two regioisomers are possible:

In both cases, undesired alkene regioisomers are possible:

For **1**, the other alkenes are tri- and tetrasubstituted, so these will likely compete strongly with the desired regioisomer. In contrast, the monosubstituted alkene from **2** should be only a minor product in this E2 reaction.

In selecting an appropriate stereoisomeric bromide, we should realize that the desired alkene stereoisomer (*E*) should come from *anti* elimination. This gives two possibilities:

SRR-2   *SSS-2*

Either of these will likely work. Given a choice, the *SRR* isomer might be preferable, because the H to be eliminated is trans to the methyl group and more sterically accessible to the base. This should help to minimize the yield of the monosubstituted alkene.

21. Several possible variations exist. What follows is based on a literature proposal (M. Rosenblum, *J. Am. Chem. Soc.* **1960**, *82*, 3796).

22. In the benzyne intermediates, just as in any alkoxybenzene, the alkoxy group tends to increase electron density by resonance at the *o* and *p* positions. Therefore, attack of a nucleophile at the *m* position is preferred.

23. First, note that cleavage of a phenol group from a mixed acetal is generally preferable to cleavage of an alkanol group, since the group that remains will stabilize the positive charge and an alkoxy group is more electron-rich than a phenoxy group. In the bicyclic case, however, stabilization by resonance with the oxygen is hampered by the constrained geometry. The resonance form shown resembles a bridgehead olefin and it appears that Bredt's rule would apply (see Section 2.3.2). But this species is different from an anti-Bredt olefin in that the O has two lone pairs, so the π bond is not really twisted. The structure does require that the π bonding involve the sp²-hybridized lone pair (shown at right), and this is not as good for two reasons: poor overlap with the C p orbital and lower energy (less available electrons). Therefore, the stabilization of the positive charge will be reduced in the bicyclic case, and hydrolysis of the monocyclic acetal will be faster.

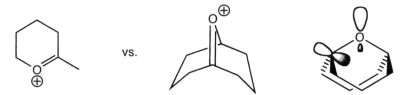

24. For most acetals, cleavage of the C–O bond is rate-limiting, and the cleavage will not occur with a significant rate unless the acetal is fully protonated. In the cycloheptatrienyl acetal, the carbocation, being aromatic (a tropylium ion), forms much more readily. Thus, the cleavage can occur concomitant with protonation, such that the protonation occurs in the rate-limiting step. The protonating acid is therefore a general acid catalyst.

25. The substrate is a thioacetal, and like an acetal, should be stable to base. The most basic site in the molecule is the methoxy group, though the pK_a of the conjugate acid should be well below 1.5. Nonetheless, protonation at the methoxy, by either a specific or general acid, could lead to loss of methanol and hydrolysis. However, heterolytic cleavage on the thiolate side is favorable without prior acid-base steps, since both the cation and anion are strongly stabilized by resonance. Acid-base steps are required to achieve the products, but these occur after the rate-determining step and therefore do not affect the rate. The mechanism shown must be favorable enough that the more typical acid-catalyzed pathway cannot compete at pH's above 1.5.

26. As described in Section 10.4.2, *anti* addition of hydrogen halides is explained by a mechanism that involves the alkene and two molecules of HX:

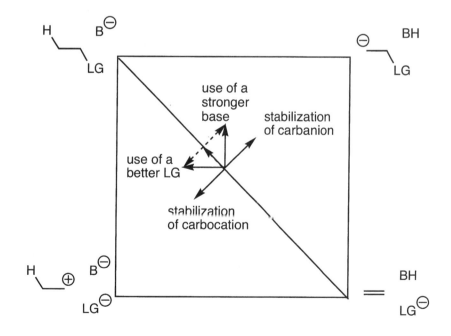

The complex formation is presumably a fast equilibrium, so the second equivalent of HBr is involved in the rate-determining step. (S is solvent and does not appear in the rate law.)

In contrast, *syn* addition is explained by reaction of the alkene with one molecule of HBr:

27. The More O'Ferrall/Jencks plot from Figure 10.11 is reproduced below, except that it shows the effects on the transition state of stabilization of the carbocation (lowering energy of the lower left corner), use of a better leaving group (lowering the energy of the bottom edge), stabilization of the carbanion (lowering energy of the upper right corner), and use of a stronger base (increasing energy of the left edge). In all cases, the effects noted in Table 10.7 are confirmed.

28. We can relatively easily show that the observed results are not consistent with an E2 mechanism. First, the expected rate law for E2 is

rate = $k[\text{ArCH}_2\text{CH}_2\text{NR}_3^+][\text{B}^-]$

The rate would thus not be expected to depend strongly on whether B$^-$ is OH$^-$ or OD$^-$. Also, the rate would be expected show a linear dependence on [B$^-$], not the observed saturation effect. If exchange did occur prior to elimination, producing $\text{ArCD}_2\text{CH}_2\text{NR}_3^+$ in D_2O, the expected primary isotope effect for E2 would be normal, not the inverse effect that is observed.

The expected rate law for an E1cb mechanism, drawn from Eq. 10.71, is

$$\text{rate} = \frac{k_1 k_2 [\text{ArCH}_2\text{CH}_2\text{NR}_3^+][\text{B}^-]}{k_{-1}[\text{BH}^+] + k_2}$$

As [B$^-$] increases, [BH$^+$] is also increased in order to keep the pH constant. The observed saturation effect is consistent with the rate law. At low [BH$^+$], the denominator may be dominated by k_2 ($k_2 \gg k_{-1}[\text{BH}^+]$), such that the rate shows a linear dependence on [B$^-$]. At high [BH$^+$], the denominator may be dominated by $k_{-1}[\text{BH}^+]$ ($k_{-1}[\text{BH}^+] \gg k_2$), such that the rate becomes independent of buffer concentration (since any increase in [B$^-$] in the numerator is cancelled by a similar increase in [BH$^+$] in the denominator).

One question that will impact the expected isotope effect is whether H/D exchange will occur prior to elimination. At low buffer concentration, where $k_2 \gg k_{-1}[\text{BH}^+]$, we would not expect exchange to occur. At high buffer concentration, where $k_{-1}[\text{BH}^+] \gg k_2$, exchange should occur. For simplicity, we will focus mainly on the low to intermediate buffer concentrations, where exchange can be neglected and any isotope effect will truly be a solvent effect.

The principle solvent isotope effect will be on the reprotonation step, represented by $k_{-1}[\text{BH}^+]$, where BH$^+$ is either H_2O or D_2O. This step would show a primary, normal KIE, since O–H or O–D cleavage is involved. Since this step is towards reactants (and $k_{-1}[\text{BH}^+]$ appears in the denominator), the expected KIE on the overall forward rate is large and inverse, as observed.

At low buffer concentrations, the $k_{-1}[\text{BH}^+]$ term drops out of the rate law, and the rate law simplifies to a form equivalent to the E2 rate law. Thus, no significant isotope effect should be observed at the far left of the plot. As the buffer concentration increases, $k_{-1}[\text{BH}^+]$ will begin to compete with k_2. However, in D_2O, the $k_{-1}[\text{BH}^+]$ term is smaller due to the KIE, so this competition (and eventual saturation) requires higher buffer concentrations.

29. The trend noted relates directly to the More O'Ferrall/Jencks plot in Figure 10.11. The overall effect of a better leaving group on an E2 transition state is that the extent of deprotonation will be reduced. Thus, the substrates with the poorest leaving group, F, experience the greatest deprotonation at the transition state and show selectivities that reflect the relative acidities of the protons (less-substituted CH is more acidic). For Cl, Br. and I, the decreasing importance of acidity becomes dominated by the stability of the product alkenes, with more highly substituted alkene being more stable. Even though the transition state becomes somewhat earlier for the better leaving groups, the double bond is apparently strong enough at the transition state to reflect the product isomer stabilities.

30. As the leaving group becomes more basic, the mechanism can shift from E1 towards Ei (Section 10.13.10). For an E1 mechanism, the regioselection occurs from the carbenium ion with transition states that are later and more product-like than the ones in the Ei process. Therefore, the stability difference between the alkenes with two or three substituents is more apparent at the E1 transition states, and more of the highly substituted alkenes are formed. In the Ei mechanism, the regioselection occurs at transition states that are more reactant-like, and the acidity differences between the protons is more important. Since protons on less-substituted carbon atoms are more acidic, the less-substituted alkene is formed preferentially.

31. The H next to Ph is more acidic than the H's on the CH$_2$. Elimination involving this H is therefore favored, even though it must proceed with *syn* stereochemistry.

32. The differences in products can be explained by a change in mechanism from E2 (or E2-like) for the cis isomer to E1 for the trans isomer. In the cis isomer, a favorable, *anti-periplanar* E2 mechanism is possible:

However, the trans isomer can only achieve an *anti-periplanar* E2 transition state from the less favorable, diaxial conformation, and then only involving the less acidic H away from the Ph group. While some of the 9% of the 3-phenylcyclohexene could come from this pathway, the array of products is best explained by an E1 pathway, which might be promoted by neighboring group assistance from the Ph group (only possible in this isomer). The resulting phenonium ion can open to the carbenium ion, which can then rearrange before losing H$^+$:

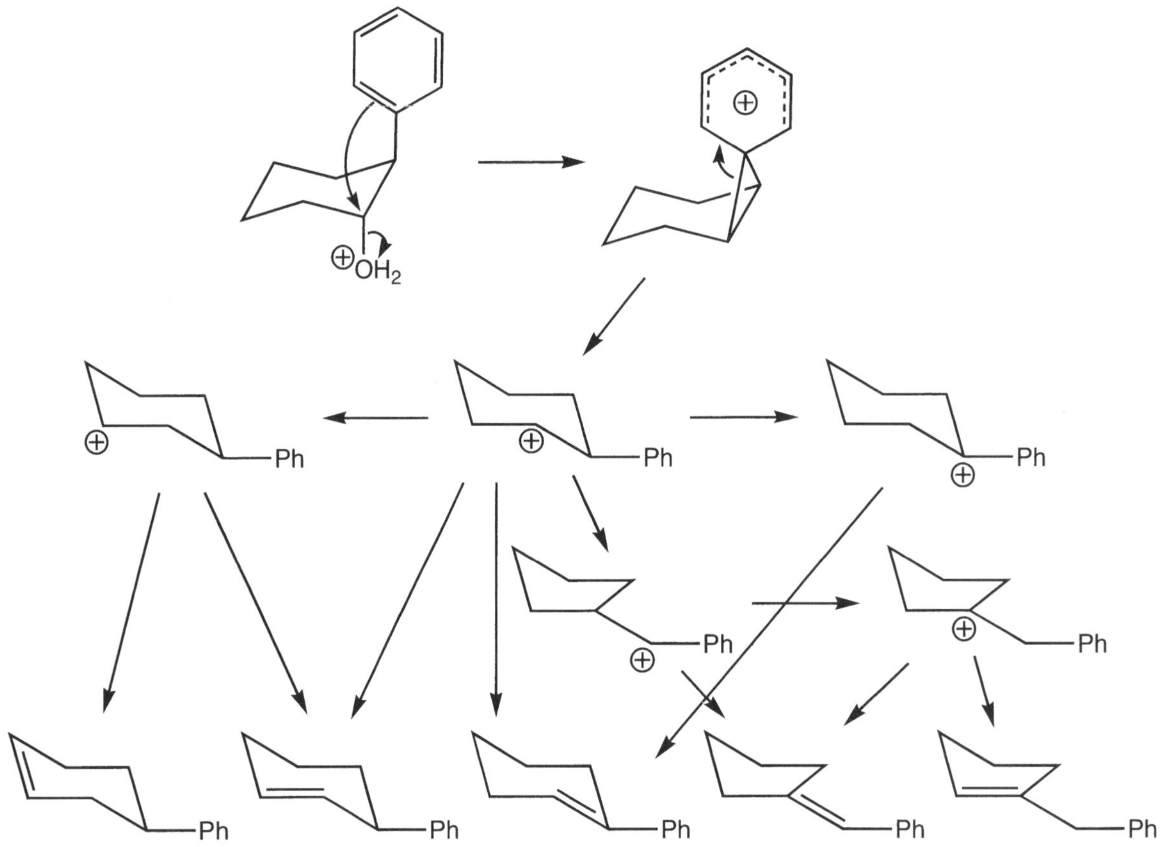

33. The two kinetically indistinguishable situations can both be represented by Figure 10.18. For straightforward general-base catalysis, the first step is rate-limiting. This step might be shown as follows to emphasize the general base:

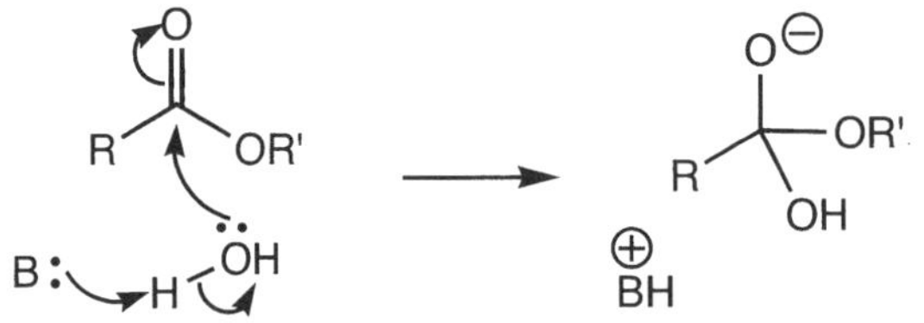

The second possibility is that the second step is rate-limiting, such that the catalysis in the first step becomes specific base catalysis. For this to be kinetically indistinguishable, the second step must be general-acid catalyzed (see Section 9.3.3 for this kinetic equivalence):

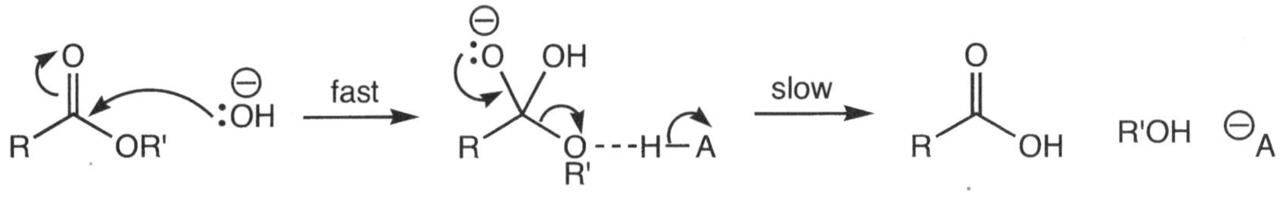

34. *t*-Butoxide is a significantly stronger base than hydroxide (conjugate acid pK_a 18 vs. 15.5), so formation of the dianion is possible (at least with a large excess of *t*-butoxide). This dianion is reactive enough to eliminate the amide anion, a poor leaving group.

35. When the acid is not strong enough, mechanism B in Figure 10.19 is not operable, and hydrolysis must occur by mechanism A. However, in very strong acid, the amide N can be protonated, allowing mechanism B to occur. Under these conditions, the slow step of mechanism A, the attack of water as a nucleophile, becomes unnecessary, and the reaction proceeds at a higher rate.

36. The cations formed from protonation of the carbonyl O are stabilized by resonance, while the cations formed from protonation of the ester O or amide N cannot be delocalized by resonance. This is shown for a protonated ester.

37. Under basic conditions, the acidity of the carboxylic acid serves to drive the coupled equilibria toward the alcohol and carboxylate. Under acidic conditions, no such acid-base equilibrium participates, and formation of the ester can be efficient.

38. Protonation at OEt and loss of EtOH would produce a cyclic oxocarbenium ion, while protonation at the O in the five-membered ring would lead to a ring-opened ion. The first process is preferable, since the developing carbenium ion has its p orbital aligned for good overlap with the aromatic π system. In contrast, the second ion is formed in a conformation in which the p orbital is rotated approximately 90° with respect to the aromatic π system. Therefore, this cation would not be stabilized by the aromatic ring until after its formation, when rotation about the C–C bond can occur.

39. The following mechanism is viable due to the relative stability of the *t*-butyl cation.

In this mechanism, neither C–O bond involving the carbonyl is broken, so ^{18}O labels in either position of the ester should remain in the carboxylic acid product. (This experiment would require care, since loss of the label due to exchange between solvent and reactant and/or product is expected, though hopefully on a slower time scale.) Another experiment would be to add a carbocation trap, such as a thiol, RSH. Formation of RS*t*Bu would be consistent with the intermediacy of *t*Bu cation.

40. The methyl groups have two effects: (1) electron donation, producing a more electron-rich aromatic ring, and (2) a large steric effect that keeps the ring and carbonyl planes roughly perpendicular. Therefore, the ring p orbitals are aligned with the C–O σ bonds, such that direct cleavage of the C–OMe bond can be assisted by this overlap. This cleavage can also be assisted by a general acid. The intermediate acyl cation is stabilized both by O and the ring.

Experiments that might help support this mechanism include LFER studies (replacing the *p*-Me with other substituents and measuring rates), isotope labeling (making the ester with ^{18}OMe and verifying that the label ends up in the MeOH), and KIE (replacing the *p*-Me–or all three methyls–with *p*-CD$_3$ and verifying that k_H/k_D is consistent with a normal hyperconjugative stabilization of the forming cation).

41. In the following equations, "I" represents the last intermediate, and the rate constant subscripts refer to either the first, second, or third step. If the first step is rate-determining, then this step will be the only one that affects the rate:

Rate = k_1[EtOH][B]

If the second step is rate-determining, then we start with the rate expression for this step:

Rate = k_2[EtO$^-$][acetone]

The first step is an acid-base equilibrium, from which we can get

$$\frac{k_1}{k_{-1}} = \frac{[EtO^-][BH^+]}{[EtOH][B]}$$

Solving for [EtO$^-$] and plugging into the rate law gives

$$Rate = \frac{k_1 k_2 [EtOH][B][acetone]}{k_{-1}[BH^+]}$$

If the third step is rate-determining, then

Rate = k_3[I][BH$^+$]

Taking the steady-state assumption for I,

k_2[EtO$^-$][acetone]$-k_{-2}$[I]$-k_3$[I][BH$^+$] = 0

$$[I] = \frac{k_2 [EtO^-][acetone]}{k_{-2} + k_3 [BH^+]}$$

Substituting into the rate law gives

$$Rate = \frac{k_2 k_3 [EtO^-][acetone][BH^+]}{k_{-2} + k_3 [BH^+]}$$

Using the acid-base equilibrium expression for [EtO$^-$] gives

$$Rate = \frac{k_1 k_2 k_3 [EtOH][B][acetone]}{k_{-1}\left(k_{-2} + k_3 [BH^+]\right)}$$

We now have three rate laws and can take advantage of the differences to determine which step is rate-determining. Upon changing [acetone], if we find that the rate does not change, then the first step must be rate-determining, since this rate law is the only one that does not contain [acetone]. We can perhaps distinguish between the second and third steps by changing [BH$^+$] while keeping [B] constant (in effect, changing the pH). The second rate law shows an inverse relationship with [BH$^+$], while the third is more complicated. At low enough [BH$^+$], the rate dependence on [BH$^+$] may vanish, if k_3[BH$^+$] becomes small relative to k_{-2}. If k_{-2} is too small for this condition to be practically reached, then the second and third rate laws would be indistinguishable. An alternative would be to keep pH constant and vary both [B] and [BH$^+$], as was done in exercise 28.

42. In the following plot, the specific-acid catalyzed path is through the lower-left corner, and the general-acid catalyzed path, for which protonation and leaving group departure occur together, is on the diagonal.

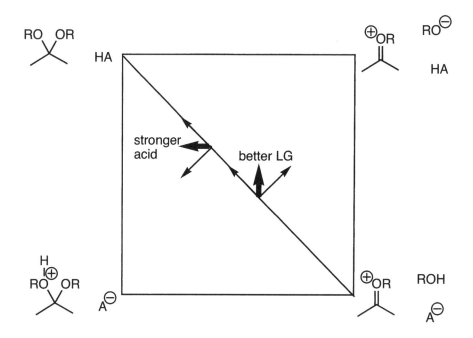

For clarity, the two changes to be considered, stronger acid and better leaving group, are shown above as occurring from different transition state positions. Making the acid stronger will stabilize the bottom edge, resulting in movement along the diagonal towards the reactants (an earlier transition state) and movement off the diagonal towards the lower left corner. The net result is that the extent of protonation at the transition state should be approximately the same, but the extent of leaving group departure will be decreased.

Making the leaving group better will stabilize the right edge, leading to an earlier transition state and one more off the diagonal toward the upper right corner. The net effect is that the extent of leaving group departure at the transition state should be similar, but the extent of protonation should be lower.

43. Like most electrophilic aromatic substitution reactions, nitration requires two mechanistic steps: addition and elimination. The addition step, making a non-aromatic sigma complex, is often rate-limiting. In the current case, however, elimination appears to be rate-limiting. This makes sense, since the high reactivity of both reactants lowers the barrier to addition. This can also lead to an apparent violation of the reactivity/selectivity principle, since the bimolecular step is both fast and reversible. In effect, regioisomeric sigma complexes can equilibrate to the most stable isomer before deprotonation. For example:

Note that in order for this scheme to give a diffusion-controlled rate, the equilibration would have to occur within the solvent cage. That is, once the reactants meet, they do not again diffuse apart.

44. The expected mechanism is addition-elimination. The stereospecificity suggests that the elimination step must be faster than rotation in the intermediate anion.

45. A.

B.

C.

46. The E2 elimination of 2-bromobutane is potentially stereoselective, because stereoisomeric products are possible (*E* and *Z*). In fact, a preference for the *E* isomer is observed, indeed making the reaction stereoselective.

Since stereoisomeric forms of the reactant also exist, the reaction would be stereospecific if the product selectivity were dependent on the reactant isomer. However, we know that this is not possible in this case, since the reactants are enantiomers and the transition states from the reactants are also enantiomeric, leading to achiral products. Therefore, any energetic preference for formation of the *E* isomer would be necessarily identical starting from either enantiomer. The reaction is therefore not stereospecific.

47. This elimination, involving a strong base and a non-acidic substrate, should proceed through an E2 mechanism, for which an *anti-periplanar* arrangement of departing groups is normally observed. This suggests that the *Z* product is the one that should be deuterated:

48. This reaction can be explained as two electrophilic additions to alkenes, followed by an intramolecular Friedel-Crafts alkylation:

(Though not shown, the source of the proton in the first step is presumably H_2SO_4 or H_3O^+, and the base for the final step is presumably HSO_4^- or H_2O.)

49. In order for a nucleophilic aromatic substitution reaction to proceed with an addition-elimination mechanism, an electron-withdrawing group, such as nitro, is generally required. The five-membered ring of this polycyclic system can serve as an electron-withdrawing group, since this ring gains aromaticity as a cyclopentadienide:

1-Chloronaphthalene is unreactive because it lacks the five-membered ring.

50. As usual, this hydroboration should go with anti-Markovnikov regiochemistry:

51. This is a free radical chain reaction. Initiation steps:

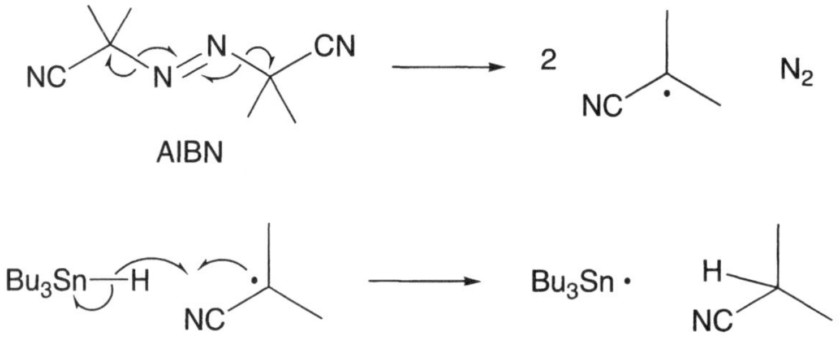

As always in chain reactions, the propagation steps should add up to give the overall reaction. Propagation steps:

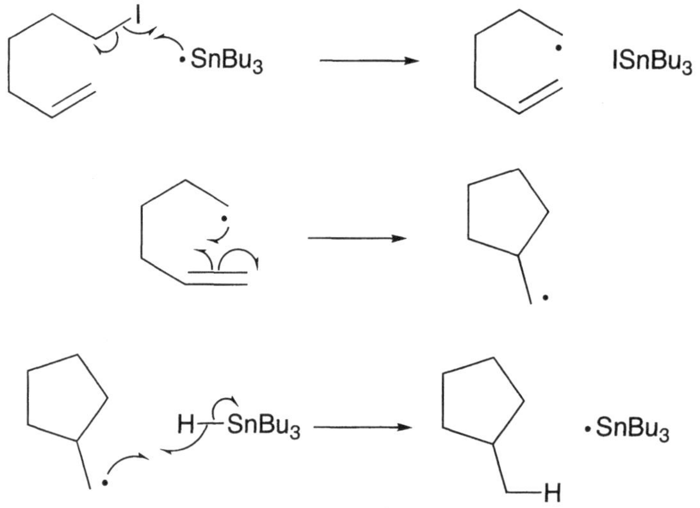

Since 5-hexenyl radical is known to cyclize selectively to cyclopentylmethyl radical (Eq. 8.73), methylcyclopentane is the expected product. The very similar ketone case gives the opposite selectivity. A possible reason for this is that the angle at the carbonyl carbon is larger (near 120°) than that at an sp^3 carbon (near 109.5°), so the closure to the smaller five-membered ring is less favorable in this case.

11

Organic Reaction Mechanisms Part 2: Substitutions at Aliphatic Centers and Thermal Isomerizations/Rearrangements

S O L U T I O N S   T O   E X E R C I S E S

1. Ion pairing with Li$^+$ reduces the nucleophilicity of the halide ions. Since Li$^+$ is small and hard, it interacts more strongly with Cl$^-$ than with the larger halides. Tetrabutylammonium, having its charge shielded by the four butyl groups, is much less able to form ion pairs. With this counter ion, Cl$^-$ is better able to compete and may be more nucleophilic due to its smaller size (more focused charge).

2. The C–H bending vibration could loosen on going to the S$_N$2 transition state as the C goes from sp^3 to sp^2-like. Conversely, one might expect a stiffening of this vibration, since the C goes from 4-coordinate to 5-coordinate. However, the distances to the Nu and LG are much longer than for normal bonds, so for the purposes of predicting the KIE, it is unclear whether the transition state is better described as 3- or 5-coordinate. The two effects counteract each other, but apparently the sp^2-like character is the more important, leading to a small but normal secondary effect.

3. At first glance, we might expect the allylic tosylate to undergo faster solvolysis due to resonance stabilization of the carbenium ion. However, the adamantyl ring system forces the alkene π orbital to be orthogonal to the empty p orbital. In the absence of resonance stabilization, the electron withdrawing character of the sp^2 C destabilizes the carbenium ion and slows the solvolysis. Thus, the first tosylate is faster.

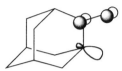

4. The scrambling occurs by dissociation to a contact ion pair (allyl cation + Cl⁻) and therefore is not accelerated by added chloride. It must also be true that the reaction of allyl cation with solvent is faster than formation of a new contact ion pair involving external Cl⁻, even at high [Cl⁻]. Otherwise, external Cl⁻ would increase the rate of return to reactant (with scrambling) relative to solvolysis.

5. In the absence of definitive information, we can presume that the transition state for the starting conditions are on the diagonal. For clarity, the two changes are taken from different starting points. In a, the change from azide to phenoxide is a change to a weaker nucleophile, corresponding to a lowering in energy of the top edge of the More O'Ferrall/Jencks plot. The net result is that the extent of leaving group departure is increased at the transition state, while the extent of nucleophilic attack is roughly unchanged. In b, the leaving group gets better as Cl⁻ is changed to I⁻, corresponding to a lowering in energy of the right edge. The net result is that the extent of nucleophilic attack is decreased at the transition state, while the extent of leaving group departure is roughly unchanged.

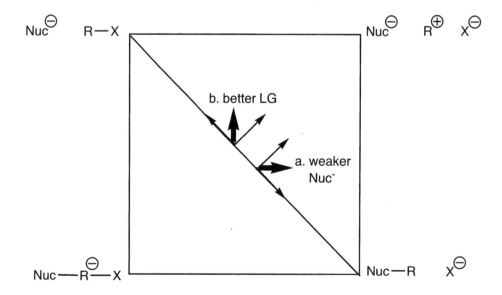

6. In the first compound the C–H(D) bond is *anti-periplanar* with the departing brosylate in the dominant chair conformation, such that this bond has a strong hyperconjugative interaction with the forming carbenium ion. As hyperconjugation weakens the C–H(D) bond, a normal secondary KIE is expected. In the second compound, the C–H(D) bond is gauche to the

departing brosylate, leading to weaker hyperconjugation at the transition state and a lower KIE.

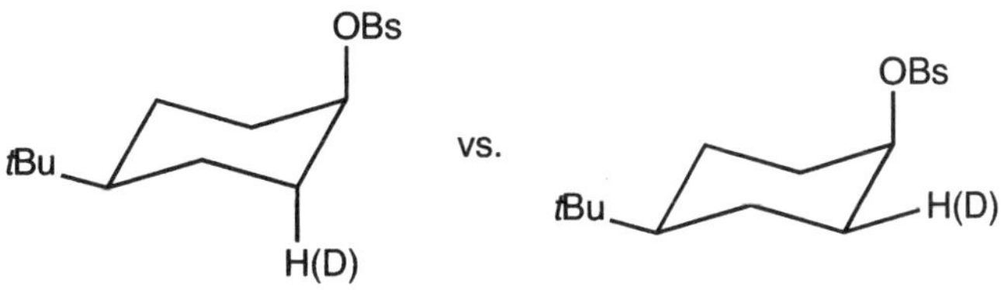

7. The second one is faster. The ortho Me groups cause the aromatic ring to rotate out of the alkene plane, such that its π orbitals align with the C–Br bond. In this conformation, the charge of the forming carbenium ion can be delocalized into the ring by resonance, lowering the energy of the transition state. In the first compound, the planar conformation, in which no such resonance is possible, is preferred.

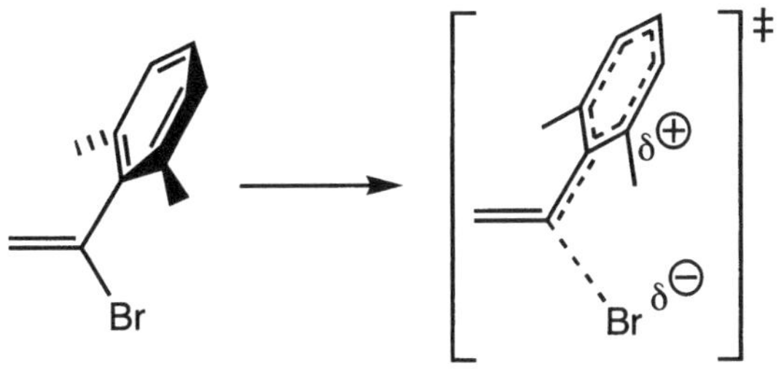

8. This substitution reaction, with an unusually poor leaving group, can be explained by an $S_{RN}1$ mechanism. Initiation requires electron transfer (the large alkyl group is represented by R):

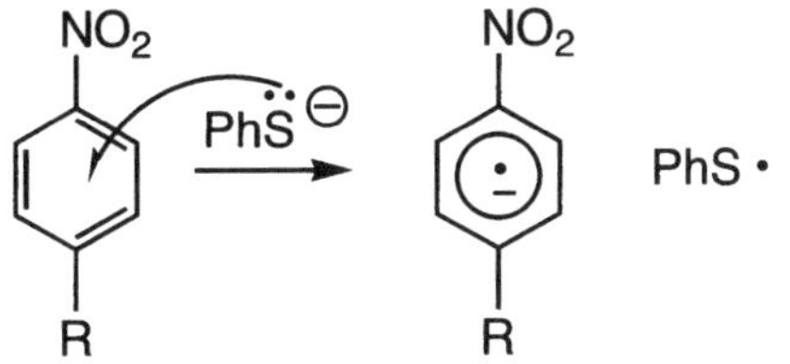

Propagation steps:

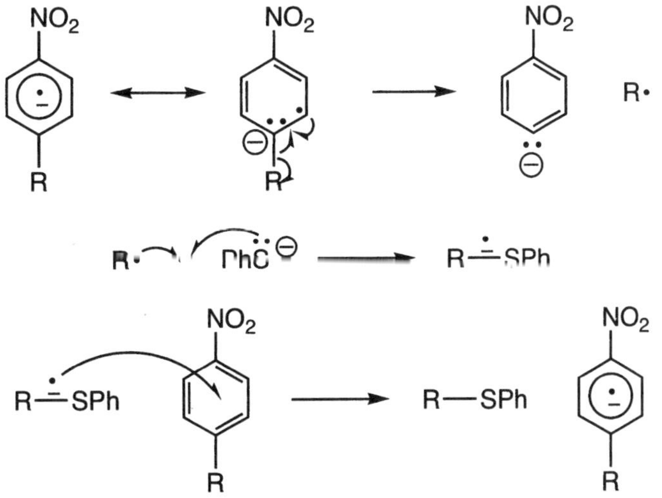

Various termination steps are possible. Also, the $C_6H_4NO_2^-$ anion will likely be rapidly protonated or otherwise react further.

9. Since the nitrophenyl ring is less electron-rich, it will be less effective at delocalizing the charge of the carbocationic intermediate. Therefore, more positive charge will remain on the benzylic carbon, increasing the importance of the hyperconjugative interaction with the C–H(D) bonds. The weakening of these bonds leads to a normal KIE and a stronger one in the nitrophenyl compound. This follows the rule of increasing electron demand (Chapter 8, exercise 25).

10. Racemization can occur within the contact (or solvent-separated) ion pair through rotation of the cation with respect to the anion before recapture:

Alternatively, the cation can be captured by a different X⁻ (which may be radiolabeled):

If racemization is faster than label incorporation into the reactant, then the first mechanism is playing a significant role. If label incorporation into the reactant is observed, then the second mechanism is also playing a role. The relative importance of the mechanisms can be determined from the relative rates of racemization and label incorporation.

11. The following reaction scheme shows the isotopic scrambling:

Starting with the isomer on the left, bond cleavage produces an ion pair that can either reform the starting isomer or the other isomer. The likelihood of either bond formation should be equal, presuming that any KIE will be small enough that it can be neglected and that the reorganization necessary is so small that the contact ion pair retains no preference for the original reactant. In other words, either bond formation is described by the same rate constant, k_{-1}. The first step is clearly rate-determining, and it is thus k_1 that can be obtained from the scrambling rate (isomerization rate). Since each cleavage has a 50% chance of forming the other isomer, the initial scrambling rate is equal to $0.5 \times k_1$[starting isomer].

12. The SET mechanism would produce 5-hexenyl, which can cyclize to cyclopentylmethyl, leading to a cyclopentylmethyl product:

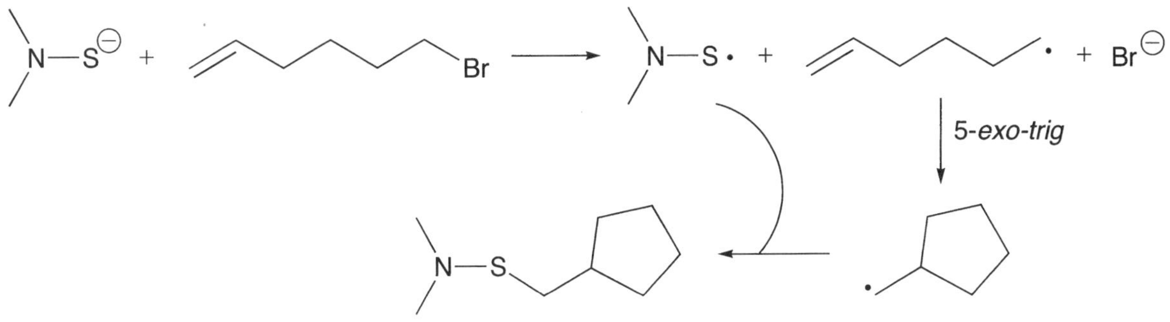

The cyclization step does take time ($k = 1 \times 10^5$ s^{-1} at 25°C, according to Section 8.8.8), so the rate of the coupling reaction is important in determining whether the cyclized or uncyclized (5-hexenyl) product is formed. A change in solvent can affect the coupling rate by making the solvent cage stronger or weaker, for example. The addition of the nonpolar pentane might be expected to weaken the solvent cage, better allowing the radicals to escape and giving the 5-hexenyl the time needed for cyclization. An alternative explanation for the solvent effect is that the more polar solvent (THF) leads to weaker ion-pairing between Na$^+$ and Me$_2$NS$^-$, such that the anion is more nucleophilic, favoring S$_N$2.

13. The π bond in the first compound is well placed for neighboring-group participation:

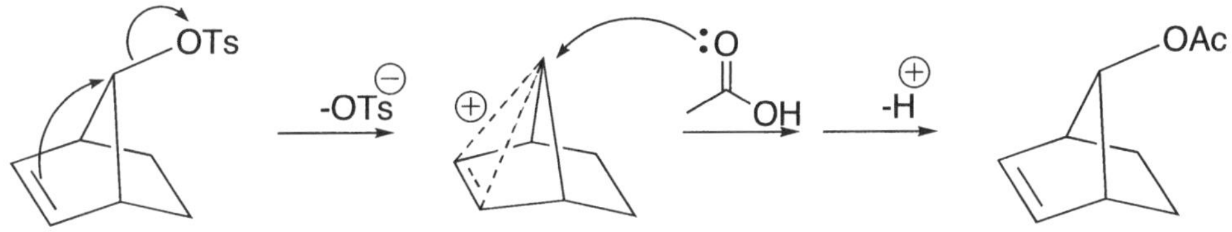

Thus proceeding through a nonclassical carbocation, this substrate undergoes much faster solvolysis. Backside displacement of the π bond by the nucleophile should result in overall retention. For the more symmetric ion from the saturated tosylate, discernment of the reaction stereochemistry would require isotope labeling. In this case, an S$_N$1 reaction involving a nonstabilized carbenium ion should go with scrambling of stereochemistry, though some preference for inversion might be observed due to ion pairing.

14. At first, this plot seems to violate the reactivity-selectivity principle in that the most reactive substrates are also the most selective. However, the reactions under study are S$_N$1 solvolysis reactions, and the reactivity and selectivity portrayed in the graph relate to different species. The reactivity (y-axis) relates to the loss of the leaving group to form a cation, but the selectivity (x-axis) is determined by the subsequent reaction of the cation with either the water solvent or added azide. The more reactive cations correspond to the less reactive alkyl chlorides and vice versa, so this plot is in fact consistent with the reactivity-selectivity principle. For example, the least reactive cation, Ph$_3$C$^+$, is also the most selective.

There are a number of cation stabilizing factors represented in the sixteen alkyl chlorides in this plot. The cations are listed below in order of the rate of the S$_N$1 reaction, in other words from least reactive to most reactive cation.

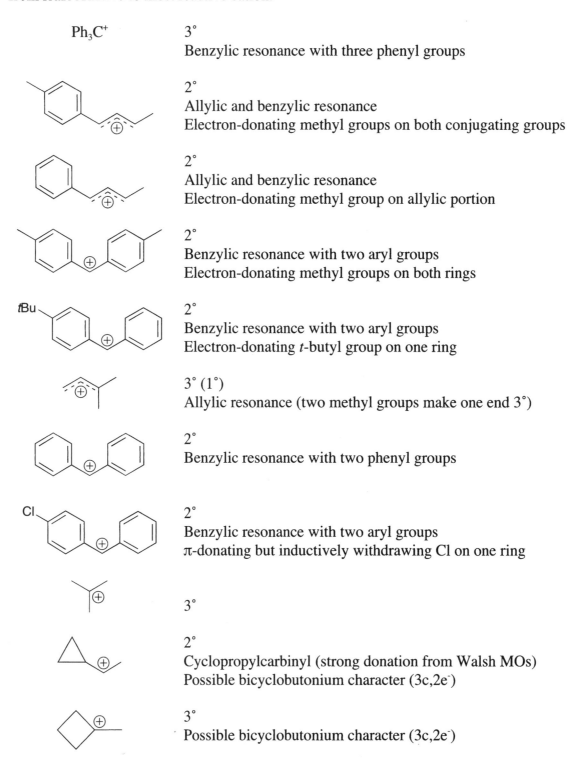

Ph$_3$C$^+$

3°
Benzylic resonance with three phenyl groups

2°
Allylic and benzylic resonance
Electron-donating methyl groups on both conjugating groups

2°
Allylic and benzylic resonance
Electron-donating methyl group on allylic portion

2°
Benzylic resonance with two aryl groups
Electron-donating methyl groups on both rings

2°
Benzylic resonance with two aryl groups
Electron-donating *t*-butyl group on one ring

3° (1°)
Allylic resonance (two methyl groups make one end 3°)

2°
Benzylic resonance with two phenyl groups

2°
Benzylic resonance with two aryl groups
π-donating but inductively withdrawing Cl on one ring

3°

2°
Cyclopropylcarbinyl (strong donation from Walsh MOs)
Possible bicyclobutonium character (3c,2e$^-$)

3°
Possible bicyclobutonium character (3c,2e$^-$)

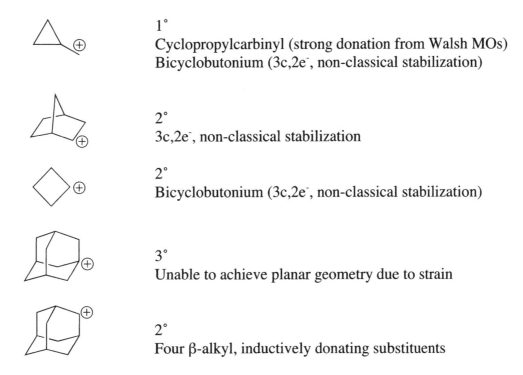

	1°
	Cyclopropylcarbinyl (strong donation from Walsh MOs)
	Bicyclobutonium (3c,2e⁻, non-classical stabilization)

1°
Cyclopropylcarbinyl (strong donation from Walsh MOs)
Bicyclobutonium (3c,2e⁻, non-classical stabilization)

2°
3c,2e⁻, non-classical stabilization

2°
Bicyclobutonium (3c,2e⁻, non-classical stabilization)

3°
Unable to achieve planar geometry due to strain

2°
Four β-alkyl, inductively donating substituents

15. At –110°C, a degenerate hydride shift occurs:

At –40°C, enough thermal energy is available to form a primary ion, which can then undergo a degenerate ethyl shift (apparently possible at the same temperature):

In the above scheme, the original cationic C is marked to illustrate that C_2 of the secondary ion is exchanged with C_1 by this mechanism. With the hydride shift that occurs even at –110°, all four C's are exchanged. Likewise, all nine H's are also exchanged.

At higher temperatures, a different primary cation can be produced through a methyl shift, and this in turn can give *t*-butyl cation after a hydride shift:

16. Attack at each of the three C's involved in the 3c,2e⁻ interaction leads to a different product: homoallyl, cyclobutyl, or cyclopropyl carbinyl. As the nucleophile attacks, the two electrons of the 3c,2e⁻ interaction localize between the two C's that are not attacked.

17. We can calculate ΔH for each reaction by using the formula:

$$\Delta H = \Sigma[BDE(\text{bonds broken})] - \Sigma[BDE(\text{bonds formed})]$$

In each reaction, the same C–H bond is broken; the BDE is approximated as that for the secondary C–H in propane.

ΔH = BDE(C–H) – BDE(H–F) = 95 – 136 = -41 kcal/mol
ΔH = BDE(C–H) – BDE(H–Cl) = 95 – 103 = -8 kcal/mol
ΔH = BDE(C–H) – BDE(H–Br) = 95 – 87 = 8 kcal/mol
ΔH = BDE(C–H) – BDE(H–I) = 95 – 71 = 24 kcal/mol

The reactivity-selectivity principle says that the least selective radical would be the most reactive one: F•.

18. These solvolysis reactions would presumably follow the S_N1 mechanism. The dramatic acceleration in the 5-oxa case suggests that the O participates as a neighboring group:

In the 4-oxa case, the O is not in a good position to participate as a nucleophile. (Quick inspection of a model of cyclooctane shows that the 1 and 5 positions are very close; 1,5 transannular reactions are often observed in eight-membered rings, while 1,4 are not.) The 4-oxa O is, however, close enough to significantly decelerate the ionization by inductive withdrawal of electron density.

19. The trends show the effects of ring size on the neighboring group cyclization rate. The general reaction is shown for the alcohol case – note that the identities of the products (cyclized or not) are not important for understanding the rate effects, since the cyclization is the rate-determining step. For the alcohols (with Cl as leaving group), the ring size effect is 5>3>6>4, while for the amines (with Br as leaving group) it is 5>6>3>7>4, with the only change in order involving 3 and 6. Two competing effects determine the ring size effect: the strain incurred in forming the ring and the loss in entropy as the reactive ends are brought together. The optimum situation occurs in the formation of the slightly strained but not-too-large 5-membered rings. Smaller rings suffer added strain, but do not give up as much entropy. As seen in for the 3- and 6-membered rings, the way the competing effects balance is somewhat reaction dependent.

20. Two different cations can be formed by the action of acid on the two OH groups. In each cation, only R_1 or R_2 can migrate. Therefore, any selectivity is likely to reflect preferences in cation formation instead of or in addition to migratory aptitude. If the cation formation were fast and reversible, then the selectivity would reflect migratory aptitude (by the Curtin-Hammett Principle), but this is probably not the case. (And even if it were, the migrations are not to equivalent cationic centers, introducing more ambiguity.)

21. A two-step mechanism accomplishes the transformation with the observed isotope labeling:

Another experiment that might support this mechanism would be the measurement of the rate at different temperatures and the determination of activation parameters. Both steps shown have highly organized transition states and the reaction would be expected to have a negative $\Delta S^{\ddagger}$. (Note that a more accurate depiction of the mechanism above would make use of O lone pairs, showing five flow arrows for the first step and six for the last. For clarity in these crowded drawings, the mechanism is shown as involving only σ and π electrons.)

22. The mechanism of each of these reactions involves backside displacement of N_2 by a group (C–C bond or O lone pair) that is properly positioned. In each case, another bond (C–H or C–C) is also properly positioned, but the observed reactions are all favored due to involvement of a lone pair on O. For each reaction, the first step is shown below. In each case, subsequent loss of H^+ gives the observed product.

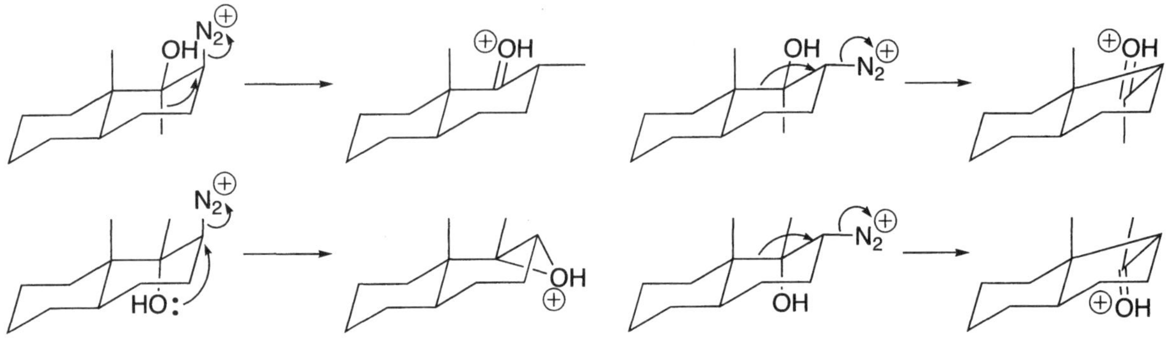

23. The mechanism is mostly the same as Scheme 11.13 with cyanide taking the place of hydroxide:

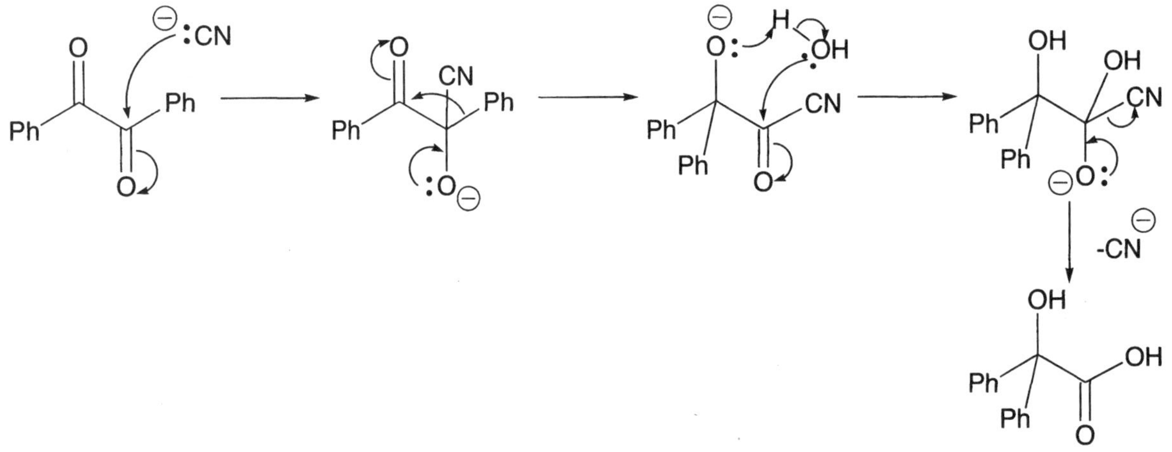

24. A.

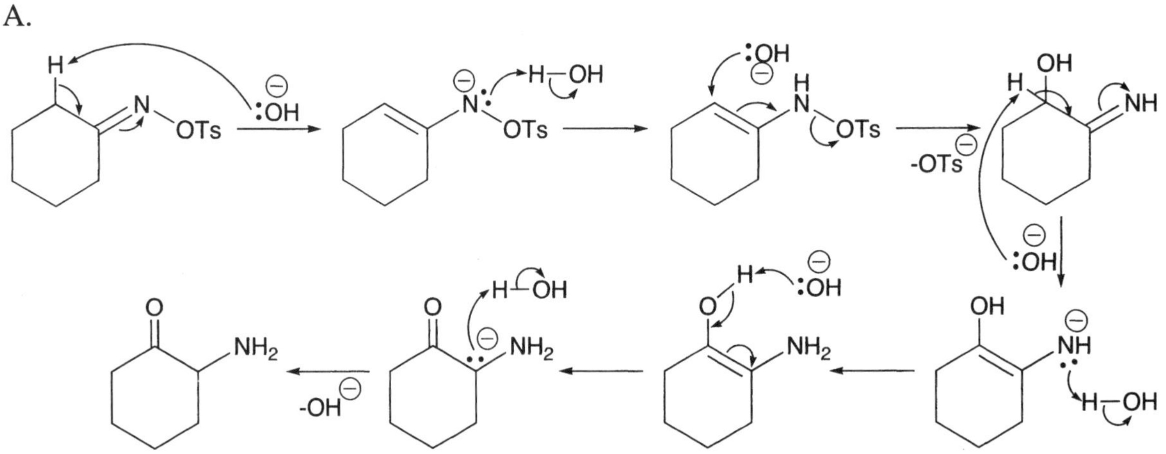

An appropriate question to ask about this reaction is, "Is the N still attached to the same C?" This mechanism says that it is, but we might envision an imine hydrolysis mechanism, for example, that would cleave this bond. We could answer this question by labeling the reactant with ^{13}C next to N. A cross-over experiment could also be done by labeling the N with ^{15}N and the ring with a Me (or D or ^{13}C), mixing with unlabeled reactant, and looking for cross-over products, which would indicate that the C–N bond is cleaved.

B.

While it's clear that N_2 is lost in the reaction, an alternative mechanism might involve cyclization and retention of the N that was originally on the end. Use of an ^{15}N label would resolve this issue.

C.

In the proposed mechanism, we might wonder whether the first or second step is rate-limiting. This could be addressed by varying p-substituents in Ar. Through a Hammett plot and direct comparisons to known substituent effects in literature reactions similar to the first two steps, this rate-limiting issue might become clear. While the aryl group does not delocalize the negative charge formed in the first step, it should have a significant effect on the electrophilicity of the carbonyl. Substituent effects on migration aptitudes are also known for similar reactions, and comparisons could be helpful.

Another potential issue is whether the benzoate product retains its original O atoms, as is predicted by the mechanism above. One can envision that attack of OH⁻ on the benzoate carbonyl in the reactant or an intermediate could lead to one benzoate O coming from hydroxide. This could be checked by using ^{18}O-labeled water (and hydroxide).

D.

If we wished to determine which step is rate-determining, we could measure the rates for derivatives with different *para* substituents and produce a Hammett plot. If the first step is rate-determining, a ρ value near zero is expected. If the second step is rate-determining, a positive ρ value is expected.

We might envision an alternative mechanism in which the C–N bond is cleaved (third step here) before the C–C bond is formed (second step here). This would create two fragments that could diffuse away from each other, and a cross-over experiment would be appropriate to investigate this possibility.

E.

The rate-determining step could be probed with a Hammett plot with substitution *para* to OH. If the first or third step is rate-determining, a positive ρ value would be expected, with the third step producing a smaller value due to the *meta* relationship with the forming charge. If the second step is rate-determining, a negative ρ value would be expected.

If the σ bond formation and cleavage steps were reversed somehow, then two fragments would be produced. The viability of this possibility could be probed with a cross-over experiment.

F.

Two questions about this reaction might come immediately to mind: (1) "Does the product OH come from the same O in the reactant?" and (2) "Does the OH end up on the ring attached to NO or N?" Clearly, isotope labeling experiments will serve to answer both question 1 (^{18}O) and question 2 (^{15}N, D, or ^{13}C). Once the identity of the substituted ring is determined, a Hammett plot for substitution on the other ring could shed light on the nature of the rate-determining step.

25. Ionization of these 3° benzylic chlorides should occur readily, and migration of the acyl group apparently occurs either at the same time as C–Cl cleavage or very shortly afterward (before C–C rotation can occur).

26. The ΔG_{298K} value given in Eq. 11.74 can be compared to ΔH computed from group increments (Chapter 2). Since the reaction is unimolecular, ΔS should be relatively small.

$$\Delta H_f^{\,\circ}(\text{cyclopropylcarbinyl}) = \Delta H_f^{\,\circ}[\bullet C-(C)(H)_2] + \Delta H_f^{\,\circ}[C-(C\bullet)(C)_2(H)]$$
$$+2\Delta H_f^{\,\circ}[C-(H)_2(C)_2] + \text{SE corr.}$$

$$= 35.82 + (-1.90) + 2(-4.93) + 27.6$$

$$= 51.66 \text{ kcal/mol}$$

$$\Delta H_f^{\,\circ}(\text{allylcarbinyl}) = \Delta H_f^{\,\circ}[\bullet C-(C)(H)_2] + \Delta H_f^{\,\circ}[C-(C\bullet)(C)(H)_2]$$
$$+\Delta H_f^{\,\circ}[C_d-(H)(C)] + \Delta H_f^{\,\circ}[C_d-(H)_2]$$

$$= 35.82 + (-4.95) + 8.59 + 6.26$$

$$= 45.72 \text{ kcal/mol}$$

$$\Delta H = 45.72 - 51.66 = -5.94 \text{ kcal/mol}$$

This value is very close to the experimental ΔG_{298K} value of –6.03 kcal/mol.

27. We can use the Arrhenius equation to calculate the rate:

$$k = Ae^{-E_a/RT} = Ae^{-(1 \text{ kcal/mol})/(0.00199 \text{ kcal/mol K})(5K)} = A(2 \times 10^{-44})$$

If the frequency factor is a normal value ($A = 10^{13}$ s^{-1}), then $k = 2 \times 10^{-31}$ s^{-1} (assuming a first order reaction – A will generally be smaller for higher order reactions). This gives $t_{1/2} = \ln 2/k = 3 \times 10^{30}$ s $= 1 \times 10^{23}$ years, a value over 10^{12} times the age of the universe! If we decide to settle for $t_{1/2} = 100$ years, we can calculate that the frequency factor would have to be 9×10^{33} s^{-1}, clearly an unrealistic value. Thus, a 1 kcal/mol barrier is essentially insurmountable at 5 K (though we should remember that tunneling is still a possibility).

28. Using group increments from Chapter 2, we can calculate the energy difference for tetramethylene and cyclobutane and compare this result to the experimental E_a of 62 kcal/mol.

$$\Delta H_f^{\,\circ}(\text{cyclobutane}) = 4\Delta H_f^{\,\circ}[C-(H)_2(C)_2] + \text{SE corr.}$$

$$= 4(-4.93) + 26.2 = 6.48 \text{ kcal/mol}$$

$$\Delta H_f^{\,\circ}(\text{tetramethylene}) = 2\Delta H_f^{\,\circ}[\bullet C-(C)(H)_2] + 2\Delta H_f^{\,\circ}[C-(C\bullet)(C)(H)_2]$$

$$= 2(35.82) + 2(-4.95) = 61.74 \text{ kcal/mol}$$

$\Delta H = 61.74 \text{ kcal/mol} - 6.48 \text{ kcal/mol} = 55.26 \text{ kcal/mol}$

This calculation says that the energy cost to form the biradical is 7 kcal/mol less than the observed activation energy. This suggests the energy diagram below, and based on this, tetramethylene is a viable intermediate. Further, it appears that there is a barrier of 7 kcal/mol for closure of the biradical (though further calculations and experiments have shown this to be an overestimation).

62 kcal/mol (from rate measurements)

55 kcal/mol (from group increments)

29. In a "general" series of nucleophiles, we should expect the C nucleophiles to be the strongest, followed by N, then O. The reasoning behind this ordering is simply electronegativity: the least electronegative, C, holds onto its electrons less tightly than does N or O. However, finding a general series of analogous nucleophiles is not a straightforward task, since isoelectronic C, N, and O nucleophiles would have different charges. Nucleophiles with the same charges would have different numbers of substituents (or bonding patterns) and different steric environments. Comparisons must be made with care. For example, Me_3C^- and Me_2N^- have different numbers of substituents, while $CH_2=C^-Me$ and Me_2N^- have the same number of substituents but different bonding patterns.

30. The stereochemical results of Eq. 11.84 are indeed consistent with conrotatory closure of the trimethylene biradicals, presuming that the biradical is formed from the diazene in a disrotatory manner. If the elimination occurs from a folded, envelope conformation, little rotation is necessary to form the planar biradical.

An alternative explanation is that the C–N bonds are broken sequentially, resulting in an intermediate diazenyl biradical. Backside displacement of N_2 from this biradical also is consistent with the preferred stereochemistry.

31. a. The disubstituted carbonyl in a ketone is thermodynamically more stable than the monosubstituted carbonyl in an aldehyde, in much the same way the increasing substitution on an alkene increases stability. Thus, the aldol equilibrium is generally more favorable for aldehydes.

b. If acetone is the solvent, a 2% impurity will represent much more than a 2% impurity with respect to the solutes.

c. At higher temperatures, the β-hydroxy ketone product will tend to undergo irreversible dehydration to an α,β-unsaturated ketone. Therefore, the amount of this product will build up beyond the 2% level, becoming an even greater impurity.

32. In the carbenium ion intermediate of an S_N1 reaction (and in the transition state leading to it), β C–H or C–D bonds are weakened by hyperconjugation. Bond-weakening leads to a normal KIE, but this effect is much smaller than a primary effect, where the bond is completely broken.

33. A likely mechanism is cleavage to the cyclopentanediyl, followed by a 1,2-H shift.

The rate-determining step of this transformation would be the bond cleavage, so the activation energy should be roughly equal to the BDE for the breaking bond. This can be calculated by using group increments. In the calculations below, the strain energy of bicyclopentane is taken from Figure 2.15. The strain energy of cyclopentanediyl is taken as equal to that of cyclopentene, taken from the same figure, since both have two sp^2 carbon atoms. The group increment $\Delta H_f^{\circ}[C-(C\bullet)_2(H)_2]$, as for trimethylene in the chapter, is taken as –4.95 kcal/mol.

$$\Delta H_f^{\circ}(\text{bicyclopentane}) = 2\Delta H_f^{\circ}[C-(H)(C)_3] + 3\Delta H_f^{\circ}[C-(H)_2(C)_2] + \text{SE corr.}$$

$$= 2(-1.90) + 3(-4.93) + (57.3)$$

$$= 38.71 \text{ kcal/mol}$$

$$\Delta H_f^{\circ}(\text{cyclopentanediyl}) = 2\Delta H_f^{\circ}[\bullet C-(C)_2(H)] + 2\Delta H_f^{\circ}[C-(C\bullet)(C)(H)_2]$$

$$+\Delta H_f^{\circ}[C-(C\bullet)_2(H)_2] + \text{SE corr.}$$

$$= 2(37.45) + 2(-4.95) + (-4.95) + 5.9$$

$$= 65.95 \text{ kcal/mol}$$

BDE = 65.95 kcal/mol – 38.71 kcal/mol = 27.24 kcal/mol

34. The C-based radical that is first formed from diethyl ether is stabilized by a 2c,3e⁻ interaction with the adjacent O atom. The radical from diisopropyl ether is even more stable, being 3°.

35. A simple mechanism with an "allyl + p" biradical explains the reaction:

One way to test for the biradical mechanism would be to run the reaction in the presence of a spin trap (Section 8.8.8) to check for the presence of trappable radicals. Alternatively, a good H-atom donor such as 1,4-cyclohexadiene or H_2Se could be added in an attempt to convert the radical centers to C–H bonds. Detection of 1- or 2-pentene in the product mixture would support the proposed mechanism. A third strategy would be to incorporate a substituent that would induce a rearrangement in the intermediate biradical. For example:

36. Cleavage of spiropentane to give a cyclopropyldicarbinyl biradical could explain stereochemical scrambling of substituted spiropentanes, as well as the conversion to methylenecyclobutane:

The second step of this sequence is just a cyclopropylcarbinyl to allylcarbinyl rearrangement, but should be especially favorable, since one of the resulting radical centers is stabilized by allylic resonance. Cleavage of the product (the reverse of the last step) can explain both positional and stereochemical scrambling in substituted methylenecyclobutanes through reclosure at the other end of the allyl radical and rotation about C–C bonds.

This reaction could be studied in a number of ways. Isotope labeling with D or ^{13}C could be done:

Labeling patterns different than the two shown would be inconsistent with the proposed mechanism, as long as care is taken to ensure that product scrambling is not interfering (which might not be possible if it is much faster than the conversion of spiropentane). The methods mentioned in exercise 35 would also apply here.

37. The migration step from Scheme 11.12 is shown below for the migration of substituted aryl groups. Since no resonance effects are normally drawn in this step, we might think that σ values, reflecting inductive effects, are the most appropriate substituent constants to use. However, a relatively poor correlation is obtained, as shown at the left below. (The migratory aptitudes in Table 11.12 are relative rates, *i.e.*, k/k_0, so the y axis is the log of these values.) Thinking a bit more carefully about the transition state (drawn at right), it seems that the aromatic π system could be involved in resonance delocalization of the positive charge, so perhaps σ⁺ values would be more appropriate. Indeed, a much better correlation is obtained for these values. (Table 8.2 gives σ⁺ values only for *para* substituents. Some σ⁺ values have been measured for *meta* substituents, and these values are generally close to σ values. So σ values were used for the *meta* substituents.)

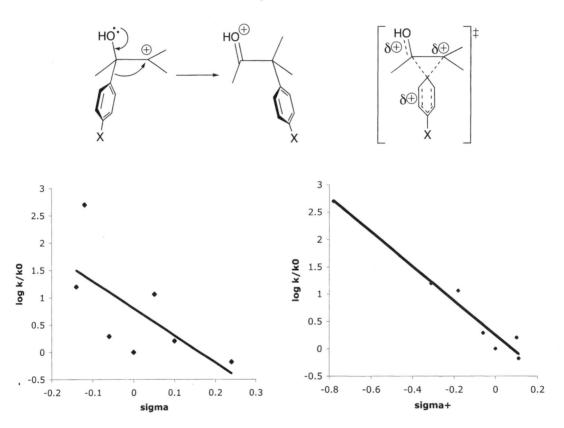

The right-hand plot has a slope of ρ = -3.2. Donating groups promote the reaction, as expected, and the magnitude of ρ is fairly high, showing that the reaction is reasonably sensitive to the substituents.

38. The largest primary isotope effects occur when the transition state is symmetric with respect to the transferring H(D). The fact that this occurs when the reactants and products are similar in energy suggests that the simple analysis that accompanies Hammond's Postulate applies well to this reaction. The reaction has an early transition state when it is exothermic and a late transition state when it is endothermic.

39. The reaction is stereospecific to a degree, since the product distribution depends at least a little bit on the reactant isomer used. In this case, the opposite ratio is necessarily observed, since the reactants are enantiomers and the entire reaction coordinates are enantiomeric. To the extent the reaction is stereoselective, it is also stereospecific. Nonetheless, when percentages are assigned, they are different – due to convention. Thus, the reaction of each enantiomer would be called 55% stereoselective, but the reaction is only 10% stereospecific. (See exercise 22 in Chapter 6.)

40. The C–D bond, with an axial orientation, is selectively removed due to its high overlap with the carbonyl π^* orbital.

41. Cleavage of one of the cyclopropane C–C bonds (the one involving both sp^3 C's) both relieves the ring strain and forms a biradical (a substituted trimethylenemethane). This biradical can then close to form a similar but more stable diene in which the exocyclic double bond is more highly substituted.

Most bond-cleavage reactions would be expected to produce increases in freedom at the transition states, giving positive $\Delta S^{\ddagger}$ values. Thus, the observed $\Delta S^{\ddagger}$ value of –6.0 eu might at first seem inconsistent with the proposed mechanism, but in this case the bond cleavage should in fact lead to a more ordered structure. In the reactant, the vinyl group can freely rotate, but this rotation becomes restricted in the biradical as the vinyl group comes into conjugation. The $\Delta H^{\ddagger}$ value of 23.8 kcal/mol is quite low for a C–C cleavage reaction; the release of strain and the introduction of extensive conjugation serve to greatly reduce the barrier.

42. Any nonclassical carbocation with a 3c,2e⁻ structure has three resonance structures, with the charge on each of the three C's. In this case, two of these structures are enantiomers, so there are two different (diastereomeric) bromides that can scrve as precursors. Note that stereochemistry is important; the Br should be placed so that the two electrons can displace it from the back side.

43. Given the pK_a values, of the two bases only the amide would serve to fully deprotonate acetone to give the enolate. Using this base with appropriate conditions (non-acidic solvent, low temperature, protection from moisture), the enolate could be prepared and then treated with the alkylating agent. Methoxide, whose conjugate acid's pK_a value is lower than that of acetone by 3 units, could at most produce a 0.1% yield of the enolate (starting with equimolar acetone and base), but this base might still be used with a different strategy. If the acid-base step can be done in the presence of the alkylating agent, then the enolate might be made to react as it forms, potentially producing a high yield of the alkylated product. Possible complications in this strategy stem from the reactivity of methoxide as a nucleophile, reacting either with the alkylating agent directly or with the carbonyl of acetone, producing a new alkoxide that can react further. But from the pK_a values alone, either base might be made to work under appropriate reaction conditions.

44. This is a multistep carbocation rearrangement that starts with protonation at the alkene:

Several points should be made. First, though all of the 1,2 shifts are potentially reversible, most of them are essentially irreversible due to the release of ring strain. Second, the four-membered rings in the reactant are all attached at spiro, sp^3 C's, so the "planes" of these rings are all perpendicular to the plane of the paper. Third, after the first 1,2 shift, which could equally well migrate either the "up" or "down" CH_2 group, each of the next three 1,2 shifts is stereoselective. In each, migration occurs to form a cis ring fusion, which is less strained than a trans ring fusion between five-membered rings. Finally, the product and most of the intermediates are chiral, but which enantiomer of the product is formed is not determined until the final deprotonation since the last intermediate is achiral. (Other products, involving even more steps, are also observed in this reaction: Fitjer, L; Quabeck, U. *Angew. Chem. Int. Ed. Engl.* **1989**, *28*, 94.)

45. Secondary substrates are generally the only ones that can follow both of the possible mechanisms. Primary substrates are limited to S_N2, avoiding the formation of unstable primary carbenium ions, and tertiary substrates are limited to S_N1 due to the steric inaccessibility of the S_N2 pathway.

46. The two carbenium ions, produced by ionization of the tosylates, are shown:

If we assume that the 7:3 ratio reflects the relative stabilities of the cations, then the energy difference is only 0.5 kcal/mol, favoring the secondary cation. While 2° ions are usually less stable than 3° ions (by 15 kcal/mol, based on the HIA values for isopropyl and *t*-butyl cations in Table 2.8), there does appear to be more ring strain in the 3° ion. Instead of all six-membered rings, this structure has one five- and one seven-membered ring. A crude estimate of the ring strain can be obtained by summing the ring strain values for cyclopentane and cycloheptane: (6.2 + 6.2) kcal/mol = 12.4 kcal/mol. This is close enough to 15 kcal/mol that we should consider it possible that these ions are nearly equally stable.

The two cation structures shown represent two of the three resonance forms for the potential carbonium ion intermediate:

47. The first rearrangement is a 1,4-phenyl shift. This seems quite reasonable, going through a bridging phenyl ring anion similar to that shown in Eq. 11.64. The product ion is more stable than the reactant, having the negative charge delocalized by resonance with the non-migrated phenyl group.

The second rearrangement is a simple 1,2-alkyl shift. Though simple to draw, such shifts do not generally occur, since the transition state is a destabilized, 3c,4e⁻ intermediate.

48. If we were to produce a three-point Hammett plot from this data, it would clearly be bent and non-linear. A curved or bent Hammett plot is usually a sign that the substituents are causing a change in the mechanism. For a simple S_N2 mechanism, the H and OMe trend is as expected; the partial positive charge on the central carbon is stabilized by delocalization into the electron-rich aromatic ring. Since the NO_2 substrate is also faster than the parent, a different mechanism must be operating for this case. One explanation is that electron transfer to the electron-poor aromatic ring occurs and that this substitution reaction follows an SET mechanism. An alternative explanation (the one given in the literature: Vitullo, V. P.; Grabowski, J.; Sridharan, S. *J. Am. Chem. Soc.* **1980**, *102*, 6463) is that all three reactions follow an S_N2 mechanism but that the tightness of the transition state increases as the substituents become less donating and more withdrawing. In the loose transition state with an OMe substituent, the partial positive charge on the central carbon is stabilized. In the tight transition state with an NO_2 substituent, the closer proximity of the nucleophile and leaving group actually lead to a partial negative charge at the central carbon, and this charge is stabilized by the electron-withdrawing substituent. These explanations are similar in that they are both based on an EWG-induced electron flow from the nucleophile to the substrate.

49. Aside from acid-base steps, this reaction requires only one additional step, though one more intermediate (a cyclopropylcarbinyl cation from opening the epoxide) could also be drawn.

50. a. An example of a good nucleophile that is also a very weak base is iodide. Its nucleophilicity is attributable to its high polarizability and its large size, which prevents effective solvation. Its low basicity is attributable to its high thermodynamic stability and the relatively weak bond in its conjugate acid, HI. An example of a poor nucleophile that is also a strong base is diisopropylamide, iPr$_2$N$^-$ (the anion of LDA). The poor nucleophilicity is attributable to its bulky iPr groups that kinetically hinder reactions with many electrophiles. The high basicity is attributable to the strong N–H bond in the conjugate acid. In explaining both cases, it's important to remember that nucleophilicity is a kinetic property, while basicity is usually taken as a thermodynamic property.

b. Nucleophilic aromatic substitution reactions often proceed readily with relatively poor leaving groups like fluoride or methoxide. In cases where the mechanism is addition-elimination, the bond to the leaving group is not cleaved until after the rate-determining step, so the rate is not strongly dependent on the identity of the leaving group – at least for reasonable leaving groups. In contrast, both major aliphatic mechanisms, S_N1 and S_N2, involve loss of the leaving group in the rate-determining step. (Aliphatic substitutions that proceed with electron transfer, SET or $S_{RN}1$, can also show less sensitivity to the identity of the leaving group.)

51. A. Though the benzylic case would undergo faster S_N2 reactions than the iPr case, it would also be more prone to reaction by an S_N1 mechanism through resonance stabilization of the cation. The iPr substrate is therefore the better choice.

B. Triflate is a much better leaving group than bromide, promoting both S_N1 and S_N2 reactions. However, bromide is the better choice, since its leaving ability is generally sufficient for S_N2 and it would lower the probability that the substrate would ionize without the assistance of a nucleophile.

C. Clearly, the better nucleophile will promote S_N2. Though both are nitrogen based, N_3^- is a much better nucleophile than Me$_3$N, due to its negative charge and its higher polarizability.

D. Though both solvents are protic, and therefore non-ideal for promoting an S_N2 mechanism, CH$_3$OH is less acidic and has less ionizing power than CF$_3$CH$_2$OH. The better choice is CH$_3$OH.

E. Since the rate law for an S_N2 reaction is rate = k[substrate][nucleophile], a higher concentration of the nucleophile should promote this mechanism over an S_N1 mechanism with a rate law of rate = k[substrate].

52. A.

B.

C.

D.

53. The molecular formula of the reactant is $C_{10}H_{17}BrO$, so the product is formed with elimination of HBr. One possibility is an E2 elimination to give an alkene, but the presence of the carbonyl with α-hydrogen atoms on each side produces two other possibilities. Indeed, the conditions are suggestive of formation of either thermodynamic (warm *t*BuOK/*t*BuOH) or kinetic (LDA, -60°C) enolates. The higher temperature and protic conditions allow equilibration of the enolates, while the lower temperature and aprotic conditions do not. Either enolate could then undergo an S_N2 reaction with the alkyl bromide, producing new five- or seven-membered rings.

The stereochemistry of the reactant is not shown, but two isomers are possible. Both isomers give the same thermodynamic enolate, so the stereochemistry of the first product is set by the S_N2 reaction. A cis ring fusion, being less strained, would be expected. In the LDA reaction, the initial stereochemistry of the reactant, cis or trans, should be preserved in the product, and the larger ring should readily accommodate either ring fusion.

54. This is a mixed aldol reaction, but there is no ambiguity concerning the roles of the ketone and aldehyde, since the enolate of the ketone is prepared before adding the aldehyde. The use of the strong base, LDA, and the low temperature will give preferentially the kinetic enolate, both in terms of regiochemistry (less substituted) and stereochemistry (*E*). As shown in Figure 11.2, the *E* enolate should favor the *erythro* (*anti*) product.

CHAPTER

12

Organotransition Metal Reaction Mechanisms and Catalysis

S O L U T I O N S   T O   E X E R C I S E S

1.

	$Ni(CO)_4$	$[(OC)_4Re]_2$ $(\mu\text{-}Cl)_2$	$(\eta^6\text{-}C_6H_6)Cr(CO)_3$	$CpFe(CO)_2\text{-}$ $(\eta^3\text{-}C_3H_5)$	$Cp_2Zr\text{-}$ $(\eta^2\text{-}Ph_2CO)$
valence electrons	Ni 10 -charge 0 4CO _8_ total 18	Re 7 -charge 0 4CO 8 Cl (covalent) 1 Cl (dative) _2_ total 18	Cr 6 -charge 0 C_6H_6 6 3CO _6_ total 18	Fe 8 -charge 0 Cp 5 2CO 4 allyl _3_ total 20	Zr 4 -charge 0 2Cp 10 Ph_2CO _2_ total 16
oxidation state	cov. L 0 charge _0_ ox. st. 0 Ni(0)	cov. L 1 charge _0_ ox. st. 1 Re(I)	cov. L 0 charge _0_ ox. st. 0 Cr(0)	cov. L 2 charge _0_ ox. st. 2 Fe(II)	cov. L 4 charge _0_ ox. st. 4 Zr(IV)

d electron count	Ni 10
	 -ox. st. _-0_
	           10
	 d^{10}
		Re 7
		 -ox. st. _-1_
		           6
		 d^6
			Cr 6
			 -ox. st. _-0_
			           6
			 d^6
				Fe 8
				 -ox. st. _-2_
				           6
				 d^6
					Zr 4
					 -ox. st. _-4_
					           0
					 d^0

Several points need to be made:

$[(OC)_4Re]_2(\mu\text{-}Cl)_2$: The structural formula given shows each Re bonded to two Cl atoms, and the bonds look the same. So why did we call one covalent and one dative? The reason is that we know that Cl has seven valence electrons and forms only one covalent bond in order to give a filled octet. If Cl already has one bond and a filled octet, it can only form a second bond by donating one of its lone pairs, forming a dative bond. (The only ways Cl could take

206

on a second covalent bond would be to accept a formal positive charge or more than eight electrons.) In a case like this, it is almost always better to preserve the symmetry of the complex, giving each Re one covalent and one dative bond from Cl:

$CpFe(CO)_2(\eta^3-C_3H_5)$: The allyl ligand is drawn as an η^3-ligand, that is with all three carbon atoms coordinated. We can count electrons most easily by considering one of the two resonance forms shown, each having one covalent bond and one datively bound alkene. The allyl therefore donates three electrons.

The total Fe valence count of 20 electrons suggests that this complex will not be stable as the structure shown. Indeed, when this complex is prepared, the allyl ligand is found to have η^1 coordination, giving an 18-electron count.

$Cp_2Zr(\eta^2-Ph_2CO)$: The structural formula suggests that the benzophenone ligand is bound to Zr by two covalent bonds, and the bookkeeping above was done for this bonding scheme. We should realize, however, that a ketone complex, like an alkene complex, has two resonance forms:

For an early, highly electropositive metal like Zr, the left (oxametallacyclopropane) structure is expected to predominate. However, let's consider how the bookkeeping would differ for the dative structure. The Zr valence electron count would not change, since 2 electrons are donated by either one dative bond or two covalent bonds. The dative structure, however, has two fewer covalent bonds, leading to an oxidation state of Zr(II) and a d^2 configuration. We can rationalize the difference by noting that back bonding from Zr to the ketone, in other words, donation of the two d electrons on Zr to the ketone, will lead to the covalent, Zr(IV) structure.

General comment: As suggested by these examples, d^6 complexes are commonly observed for metals near the middle of the transition series. In these complexes, the d_{xy}, d_{yz}, and d_{xz} orbitals, either nonbonding or involved in π interactions, are filled. For early transition metals, specifically groups 3-5, d^0 is a common configuration.

2. The CO stretching data is consistent with the Dewar-Chatt-Duncanson model: stronger donors should make the metal center more electron-rich and promote back-bonding interactions. Back-bonding strengthens the M–C bond and weakens the C–O bond. Therefore, it makes sense that as the ligand donor strength is increased on going from $P(OMe)_3$ to PPh_3 to PMe_3, the CO stretches decrease. The Me groups of PMe_3 are more donating than Ph groups, which in turn are more donating than electronegative OMe groups. (Note that in a context where these groups are bonded to a good π acceptor, like an aromatic ring, the donation order would be different: OMe > Me > Ph.)

3. Based on the order of donating ability presented in exercise 2, we might expect a different order of bond strengths for the phosphorus ligands. Since PPh_3 is a better donor than $P(OMe)_3$, it might be surprising that $P(OMe)_3$ tends to form stronger bonds. However, PPh_3 is more sterically demanding, shown by its cone angle of 145° relative to 128° for $P(OPh)_3$ (a phosphite ligand from Table 12.1 that is presumably larger than $P(OMe)_3$ – indeed, the cone angle for $P(OMe)_3$ is 107°). Therefore, the potential bonding ability of PPh_3 is attenuated by its size. It is no surprise that PMe_3, the smallest and best donor of the phosphines, forms the strongest bonds.

Though one might expect the smallest ligand to be NMe_3, due to the shorter M–N bond it actually has a larger cone angle than PMe_3 (PMe_3 118°, NMe_3 132°; for amine cone angles, see A. L. Seligson and W. C. Trogler, *J. Am. Chem. Soc.* **1991**, *113*, 2520). The most important effect, however, is the weaker donor ability of NMe_3 due to the higher electronegativity of N relative to P. Therefore, M–NMe_3 bonds are weaker even than the more sterically encumbered M–PPh_3 bonds.

4. The fact that the insertion and deinsertion are both observed shows first that the entire pathway is kinetically reversible. Further, the fact that the different directions can each be favored by changing P_{CO} shows that the thermodynamics (more specifically, the sign of ΔG) depends on P_{CO}. A reasonable mechanism is shown below (analogous to Eq. 12.42):

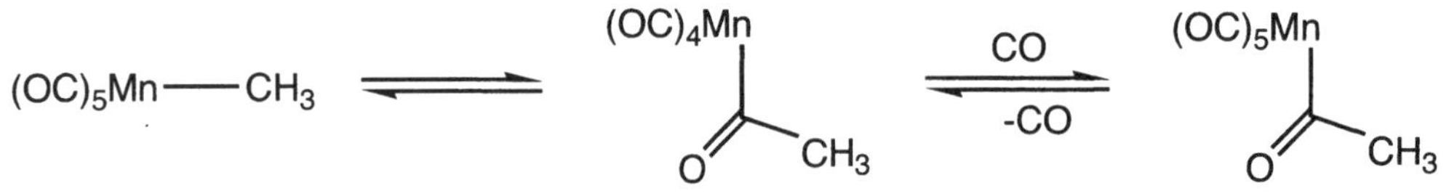

The first step of the insertion mechanism is a unimolecular migratory insertion that presumably favors the coordinatively saturated (18-electron) reactant. The second step is the one that should respond to P_{CO} according to Le Châtelier's Principle. Thus, adding CO pushes this step to the right, and removing CO pushes it to the left. A reaction coordinate diagram might look like this:

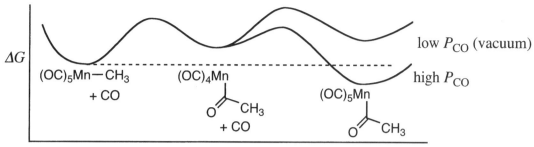

An important point about this diagram is that it is a plot of ΔG, not $\Delta G°$. On a plot of $\Delta G°$, we could not see the changes that occur as a function of P_{CO}, because $\Delta G°$ is defined for a specific P_{CO}, i.e., 1 atm. (Another technical point is that under conditions of complete vacuum, such that CO is continuously removed and $P_{CO} = 0$, equilibrium cannot be achieved and ΔG is not defined.)

5. The replacement of a CO ligand with a phosphine ligand will reduce the number of potential CO ligands that can participate in the insertion reaction (therefore slowing the reaction by a statistical factor of 0.75, presuming the phosphine to be cis to the methyl. However, a likely larger effect will be the stabilization of the 16-electron intermediate by the stronger donor ligand. Therefore, the insertion reaction should be faster for the phosphine complex.

The addition of a Lewis acid should accelerate the migratory insertion through electrophilic activation of the CO ligands. (See Section 12.2.5.)

6. The problem with RX substrates where R = ethyl, propyl, butyl, etc. is not the oxidative addition step. The problem is that these R groups have β-hydrogen atoms and can undergo β-hydride elimination immediately following the oxidative addition. This elimination step is apparently faster than the next step of the Heck reaction, insertion of the alkene, since this step would remove the R group from the metal and lead to a successful completion.

7. Triethylamine, having a lone pair on N, will first act as a ligand. Once bound to the metal (we'll use PdCl₂), the only reasonable reaction type from the chapter that we can use is β-hydride elimination. Though this elimination is a bit different than those in the chapter, involving a dative rather than a covalent ligand, it can be drawn in much the same way:

One consequence of the reactant and product ligands both being dative is that the ligand and metal each take on a formal charge as H⁻ is transferred from the ligand to the metal. (In usual β-hydride elimination, no charges are produced as a covalent alkyl ligand becomes a dative alkene ligand.) In any case, the Pd oxidation state is still +2, though we are now set up to do a reductive elimination. We can also take off the iminium ligand and the other chloride to give Pd(0) (shown without ligands for simplicity).

Though this mechanism works, it perhaps is a little strange in that most of the intermediates have a negatively charged metal center when metals more often have positive charges, if any. We can fix this by changing the order of the steps:

As a final note, it is fair to point out that the Cl⁻ ligand could leave also as the triethylamine coordinates or concerted with the β-hydride elimination.

8. We can compute $\Delta H°$ for any reaction by subtracting the sum of the product BDEs from the sum of the reactant BDEs. (If you have trouble remembering which is subtracted from which, just think about breaking a single reactant bond. The BDE is positive, as should be $\Delta H°$. So the reactant BDEs should contribute as positive.) For a general β-hydride elimination:

$\Delta H° = BDE(M–R) + BDE(C–H) + BDE(C–C) - BDE(C=C) - BDE(M–alkene) - BDE(M–H)$

$\Delta H° = (30 + 100 + 90 - 163 - 20 - 60)$ kcal/mol

$\Delta H° = -23$ kcal/mol

In addition to the exothermicity, β-hydride eliminations are often driven by entropy, since the weak M–alkene bond is readily cleaved, forming two fragments from one.

9. This mechanism requires a series of coordination and migratory insertion steps. The last step can be viewed as a rearrangement of coordination in an enolate ligand.

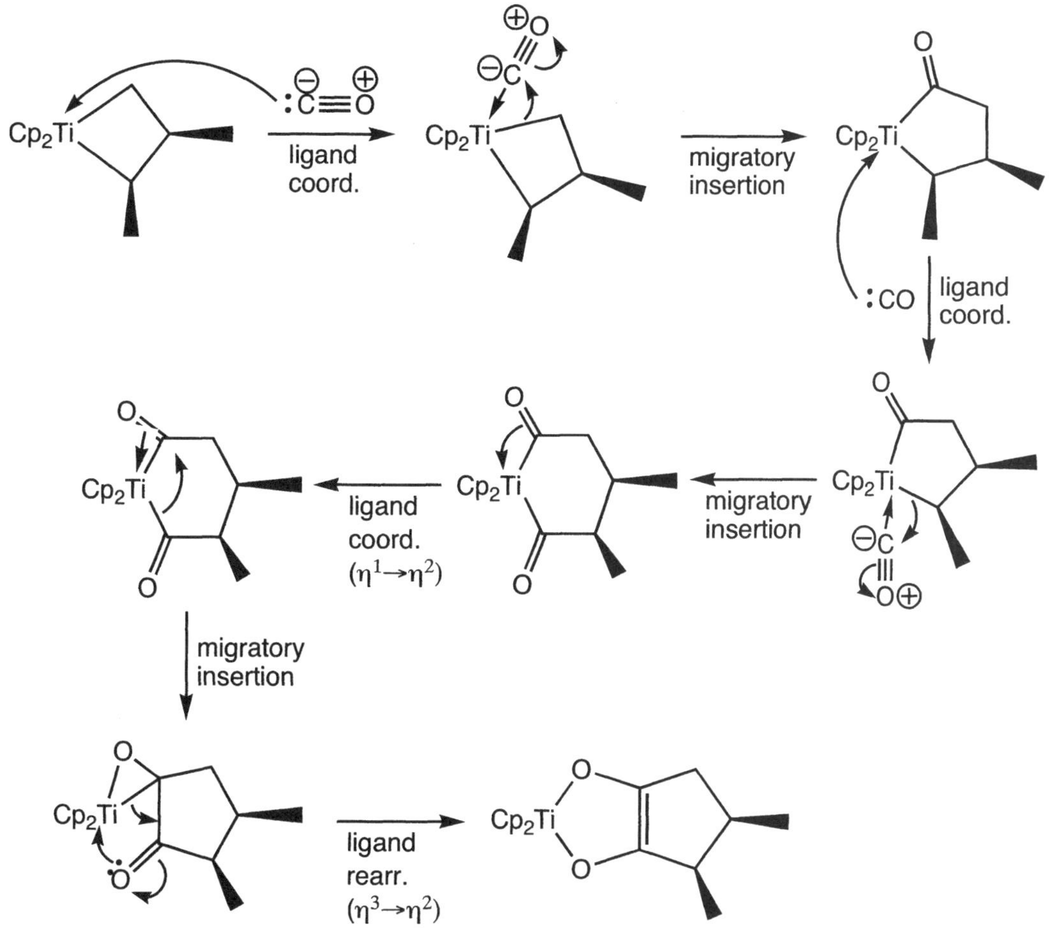

10. a. We can determine the stereochemistry at the carbon attached to the metal by making this carbon stereogenic and homochiral. Then the products arising from retention and inversion at this center will be enantiomers. For convenience, however, we might make the β-carbon also stereogenic, as shown below. This has two advantages. First, the reactant need only be diastereomerically pure – a racemic mixture will work. Second, the α retention and inversion products are diastereomers, presuming the configuration of the β-carbon will be unaffected by the reaction. This greatly expands the possibilities for analysis of the product mixture; for example, ^1H or ^2H NMR may suffice.

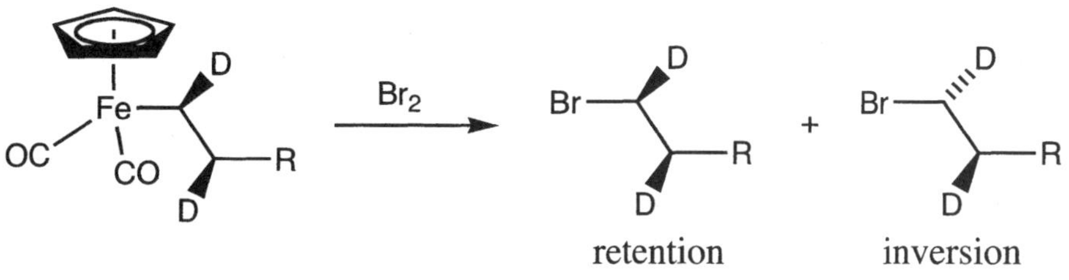

b. In short, the stereochemistry results can be explained by an S_N2 step that goes with inversion. In the case where R = Ph, the Ph group can serve as an internal nucleophile, leading ultimately to a double inversion, which is observed as retention. See Eq. 12.58 and 12.59 for the mechanism.

While the stereochemical observations are consistent with the proposed mechanism, further experiments could be done to test it. The monolabeled reactants below, with either D or ^{13}C labels, would be expected to give different label distributions. If the proposed mechanisms are correct, the case with R = tBu should give product labeled completely at the halide-bearing carbon. In contrast, the case with R = Ph should distribute the label equally between the halide- and phenyl-bearing carbons (neglecting a small expected isotope effect in the second S_N2 step). Such a result would strengthen the experimental support for these mechanisms.

11. A. The key to this exercise is recognizing that the epoxide O is a suitable leaving group, allowing the formation of an η^3-allyl complex. Reasonable mechanisms that do not employ an η^3-allyl complex can be drawn, but the formation of Pd η^3-allyl complexes from vinyl epoxides is well established.

B. After coordination of propene, nucleophilic attack of amine, and deprotonation (perhaps with more amine), oxidation of the Fe to a 17-electron center can promote the remaining steps of migratory insertion, coordination of N, and reductive elimination. Note that if the iron species were oxidized by one more electron and supplied with another CO, the starting complex would be regenerated. Thus, this reaction might possibly be made catalytic in the Fe complex.

12. A. The decrease in rate due to added PPh_3 is consistent with a loss of coordinated PPh_3 before the rate-determining step. The added PPh_3 shifts this equilibrium back toward the reactant, reducing the concentration of the first intermediate.

B. The fact that the reaction is first order in both Rh complex and PPh_3 tells us that both reactants are involved in or before the rate-determining step. The large decrease in entropy also signals a coming together of the reactants. Both point to an associative process. Since the complex has 18 electrons on Rh, it must find some other way to besides dissociation of CO to make room for the PPh_3. Since the only other ligand is a substituted cyclopentadienyl, a slippage of this ligand from η^5 to η^3 seems most likely. This could occur either before or at the same time as the PPh$_3$ coordination. (Note also that there are three possible positions for the NO$_2$ group on the η^3 ring.)

C. In correlating the reactants and products, we can see that the methyl group from methyllithium must end up as one of the methyls of the isopropyl. Therefore, we know the site of attack on the butadiene ligand in the first step. The use of deuterated acid to remove the ligand from the metal leads presumably to an electrophilic attack on an alkyl or alkenyl complex, where the Fe atom is replaced by D. The correct alkyl complex is obtained after the attack of MeLi, and we can explain the D scrambling through insertion/deinsertion steps.

D. This reaction hints at reductive elimination – the Re is reduced from Re(III) to Re(I) and the organic ligand comes off the metal without any atoms being added or subtracted. But a reductive elimination from the reactant complex would give a cyclobutane product, not a cyclopropane. A rearrangement is needed before reductive elimination, and β-hydride elimination followed by reinsertion of the alkene in the opposite direction would give the correct product, including the D labeling. This is essentially the mechanism sketched below, but we must remember that β-hydride elimination requires an open coordination site, and the Re is coordinatively saturated (has 18 electrons). We are told that CO does not dissociate, but we can open a coordination site through Cp ring slippage, as in part B.

As a final note, the PPh$_3$ appears only in the final step, consistent with the observation that the reaction is zeroth order with respect to PPh$_3$. This coordination occurs after the rate-determining step, such that PPh$_3$ does not affect the rate.

E. As in part A, the deceleration with added phosphine suggests that the mechanism requires dissociation of phosphine. The platinacyclobutane product is suggestive of an oxidative addition of a neopentyl γ C–H bond. Making this step reversible allows for the observed deuterium scrambling through multiple oxidative addition and reductive elimination steps. The mechanism below shows the formation of both neopentane and the metallacycle products as d$_0$, d$_1$, and d$_2$ versions.

13. The η^2-alkene form requires only the dative donation of two electrons from the ligand to the metal, whereas the metallacyclopropane form requires also the back-bonding donation of two electrons from the metal to the ligand. Thus, the metallacyclopropane form has the metal in a higher oxidation state by two electrons. The early transition metals are very electropositive (less electronegative than carbon), so they will tend to give up the two electrons to the alkene. In reality, a continuum of structures between the two extremes is possible, but the early metals are more likely to have structures closer to that of the metallacyclopropane (as long as they are not already d^0).

14. The cycle below includes ligand (CO) loss, reductive elimination of ArCl, oxidative addition of the acid chloride, and a migratory deinsertion.

15. A. The oxidative addition step probably proceeds by an S_N2 (or S_N2') mechanism, given the high S_N2 reactivity of allyl chloride. The reaction could reasonably be associative or dissociative, and both possibilities are shown. In the dissociative pathway, coordination of the alkene before oxidative addition and η^3-allyl coordination afterwards are drawn, though the actual mechanism does not necessarily include these. In the associative pathway, coordination of the alkene would not be expected as it would produce a 20-electron complex, and η^3-allyl coordination would occur only after CO loss for the same reason.

Associative:

Dissociative:

B. Since Br$_2$ is an electrophilic reagent, an associative mechanism might be most reasonable. (A reactive reagent is more likely to attack than to wait.) The stereochemistry is determined only in the last step, and either cis or trans isomers would be possible. The cis is favored, because the carbonyl groups, which are π-acceptors, prefer to be trans to π-donors like Br rather than to each other.

C. After CO loss, oxidative addition of H$_2$ should be concerted, giving a cis complex.

D. The trace O$_2$ serves to initiate a radical chain reaction (by reaction with either reactant or an impurity).

Initiation:

Propagation:

Termination:

Other termination steps are possible. The product stereochemistry is scrambled at the C attached to Ir and the addition is trans at Ir.

16. The product arises from an oxidative addition reaction at the square-planar, 16-electron Ir center. The loss of stereochemistry is suggestive of a radical mechanism. The C–Br bond is much more reactive than the C–F bond, and the configuration at the F-containing C is retained in the product. The stereochemistry at Ir would be the result of a trans addition.

CHAPTER

13

Organic Polymer and Materials Chemistry

SOLUTIONS TO EXERCISES

1. The G0 dendrimer is just the core, tetraaminoadamantane. We generate the G1 dendrimer by attaching two aminopropyl groups to each amine:

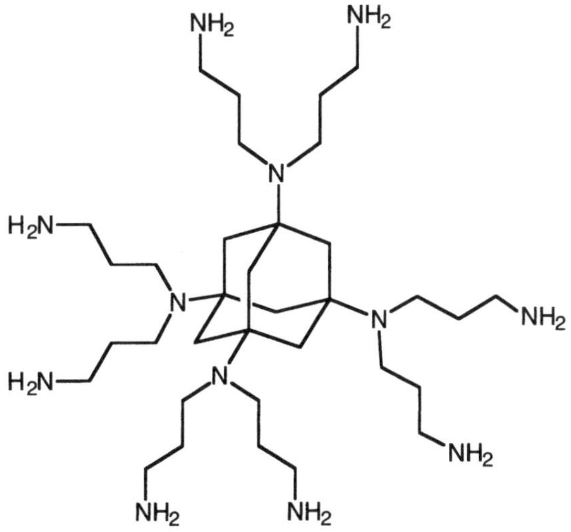

Repeating the process gives the G2 dendrimer:

The number of exposed amine groups (*i.e.*, the NH_2 groups) is 4 for G0, 8 for G1, and 16 for G2. Clearly, the number doubles with each generation, so G3 would have 32, G4 would have 64, and G5 would have 128. (Good thing we didn't have to draw that one!)

2. The overall yield will be $(0.995)^n$, where n is the number of linkages. We can see from Figure 13.4A that starting with a diaminobutane core, 4 linkages are required for G1 and 8 more for G2. The number of new linkages doubles for each successive generation. So the total number of linkages is tabulated below for up to G5.

Generation	# new linkages	total linkages (n)	Yield (% intact)
G1	4	4	98
G2	8	4 + 8 = 12	94
G3	16	12 + 16 = 28	87
G4	32	28 + 32 = 60	74
G5	64	60 + 64 = 124	54

Therefore, the yield of the intact G5 dendrimer is $(0.995)^{124} = 54\%$. In practice, the sample obtained would likely have a mass representing close to 100% yield, but almost half of the dendrimer molecules in the sample would have defects.

3. Polymers, like other substances, have higher phase transition temperatures if their attractive intermolecular forces are greater. The intermolecular forces are greater in PVC than in polypropylene for two reasons. The bond dipoles associated with the polar C–Cl bonds give rise to attractive dipole-dipole forces. Also, the higher polarizability of Cl leads to attractive London dispersion forces.

4. A polypropylene pentad is described by four diad relationships, each either *m* or *r*, giving a designation such as *mmmr*. Given four descriptors with two possible values for each, we might think there should be $2^4 = 16$ possibilities. Since this is more than the 10 expected, we should consider whether some of the 16 might be equivalent. What about *mmmr* and *rmmm*?

   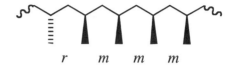

We can see that these pentads are enantiomeric and therefore would not be distinguished by ^{13}C NMR. We should, however, realize that there is another way we could have drawn *rmmm*:

   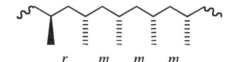

This, indeed, is identical to the *mmmr* structure at the left, being reoriented 180° about a vertical axis. Therefore, *mmmr* and *rmmm* are identical designations, either of which represents the same pair of enantiomers. Considering other possibilities of end-to-end symmetry, we find a total of six duplicates in our set of 16, leading to the anticipated 10 pentads:

mmmm	*mmmr*	*mmrm*	*mmrr*	*mrmr*	*mrrm*	*mrrr*	*rmmr*	*rmrr*	*rrrr*
	or	or	or	or		or		or	
	rmmm	*mrmm*	*rrmm*	*rmrm*		*rrrm*		*rrmr*	

GOING DEEPER

A reasonable question to ask is, "Since analysis of diads would tell us the difference between isotactic and syndiotactic and would even give us a measure of the purity, why would we want to analyze for polypropylene pentads?" The answer is that the pentad distribution gives us a much better picture of the tacticity defects that are present and may even lead to better mechanistic understanding.

For example, consider a case where we are using a chiral catalyst that produces isotactic polypropylene. Suppose that diad analysis showed that 98% of the diads were *m* and 2% were *r*. This tells us that our catalyst is fairly selective. But we might then wonder exactly how the defects come about. One possibility is that an occasional monomer is incorporated in the wrong orientation with respect to both of its neighbors. Another possibility is that once the wrong monomer is incorporated, subsequent polymerization is isotactic from that point, producing a string of the opposite stereochemistry. This could occur if the catalyst occasionally switches from one enantiomer to the other, therefore switching the preferred orientation of monomer. The diad analysis gives us no information that would allow us to distinguish these possibilities. However, the analysis of pentads would. While both mechanisms would give rise to *mmmr* pentads, only the faulty monomer orientation pathway would lead to *mmrr* and *mrrm* pentads. On the other hand, the observation of significant percentages of *mmrm* pentads would suggest catalyst isomerization.

5. In order to produce polymer samples with well-defined molecular weights, we need a method that gives low polydispersities. Indeed, ring-opening metathesis polymerization (ROMP) with some catalysts is a living polymerization, giving very low PDIs. Clearly, we cannot make polyethylene directly by ROMP, since any polymer from ROMP would have double bonds. However, we could make an unsaturated polymer that would give polyethylene after hydrogenation. Possible monomers would include cyclobutene, which has plenty of strain to drive the polymerization, and 1,5-cyclooctadiene, which, as the text mentioned, can be polymerized by the newer catalysts in spite of its low strain energy.

Note that the polyethylene obtained from either of these monomers is chemically the same, even though the drawings look different.

6. From the hint given, the retrosynthetic analysis is clear:

polyethylene glycol   MDI   hydrazine   MDI

In assembling the pieces, we must take some care with the order of steps, stoichiometry, and order of addition to ensure that the desired linkages occur cleanly. For example, if we add 2 equivalents MDI to 1 equivalent of hydrazine, the desired 2:1 product might predominate but would likely be contaminated with significant amounts of 3:2 and other adducts. Separation of these products would likely be difficult. With some thought, we can design a synthesis that avoids these problems.

Spandex

The polymer of ethylene oxide can be prepared first by anionic (or cationic) polymerization. Adding the polymer to 2 equivalents of MDI should give the diisocyanate shown. Note that the order of addition ensures that MDI is in excess until the addition is complete, disfavoring the undesired reaction of MDI with two polymer chains. To obtain the highest purity, an excess of MDI could be used rather than just 2 equivalents. The excess MDI would be easily separated by washing the polymeric product with a solvent that does not dissolve the polymer. (Note that this is one reason to start with the polymer rather than with hydrazine – ease of purification of the intermediate product.) Treatment of the polymeric diisocyanate with 1 equivalent of hydrazine would give the Spandex polymer. Precise control of 1:1 stoichiometry is very important in this step if high molecular weight polymer is desired.

7. The two different pathways are depicted below. (See Figure 13.20 for electron flow for the top pathway.)

A clear difference between the two mechanisms is that the first involves a concerted reaction, while the second involves cleavage followed by intermolecular steps. This situation always suggests the possibility of cross-over experiments. If we were to do this reaction in the presence of added $(CD_3)_3SiF$, the deuterated silyl groups would only be incorporated through the second mechanism. If the first mechanism is operative, the silyl group from the reactant (unlabeled) should be found in the product.

Other experiments that would be sure to distinguish these mechanisms are hard to find. The initial attack by F^- is likely the rate-determining step, meaning that kinetic methods (including isotope effects and LFER) are probably of no help. We could attempt to trap one or both of the intermediate enolate ions of the second mechanism, but even if successful we might find it difficult to argue that the same products could not arise from trapping of the hypervalent silicon intermediates that appear in both mechanisms.

8. Though the structure of the poly(methyl methacrylate) formed would be the same with respect to the monomer units and the connections between monomers, the polymers would differ in some respects. First, the end groups would be different. Second, since the free radicals experience termination and the GTP polymerization is living, one can expect a much lower polydispersity for the polymer from GTP. Control of molecular weight in the GTP method should be fairly simple, with MW ≈ (mol monomer)/(mol catalyst). In contrast, the molecular weight of the polymer from free radical polymerization would be much more complicated to control, since it depends on the relative rates of initiation, propagation, and termination (primarily the latter two), all of which can vary as polymerization proceeds.

9. The two reactions are shown:

One difference in structure concerns double bond stereochemistry. The polymer from cyclooctadiene should have at least 50% cis stereochemistry, since half of the double bonds remain from the reactant (presuming that metathesis at the acyclic double bonds of the polymer is slow). The stereochemistry at the other half of the double bonds, and all of the double bonds in the ADMET polymer, is determined by the stereoselectivity of the metathesis reaction (frequently favoring trans).

The molecular weight distribution will likely be very different. The molecular weight of the ROMP polymer, being an addition polymer, will likely be much higher than that of the ADMET polymer, which is a condensation polymer. The degree of polymerization of the living ROMP reaction is easily controlled by adjusting the monomer/initiator ratio, while the degree of polymerization of ADMET is limited by the yield of the individual metathesis steps (Eq. 13.10).

10. We will consider the copolymerizations one-at-a-time. The tables show the two types of radicals and monomers along with the reactivity ratio (r) for each. If $r > 1$, the radical prefers to react with a monomer of its own type, and if $r < 1$, it prefers the other monomer.

Radical	Monomer	r
		1.22
		0.15

The r values show that methyl methacrylate (MMA) is preferred by both radicals, though the preference is much greater for the acrylonitrile-derived radical. (The reciprocal of 0.15 is 6.7, showing 6.7/1.22 = 5.5, or a greater than 5-fold stronger preference.) Considering radical stabilities, the carbomethoxy and cyano groups are both electron withdrawing groups, a destabilizing effect for a radical center that is already electron-deficient, though both allow some resonance delocalization of the unpaired spin. Cyano is the stronger electron withdrawing group, based upon Hammett σ values. This effect, together with the extra methyl group in the MMA-derived radical, serves to make the MMA radical more stable.

Thus, we can understand why either radical would prefer MMA, leading to a more stable radical.

The greater preference exhibited by the acrylonitrile radical can be understood as well. This radical is more electron deficient than the other, so it will show an even greater attraction to the less electron-deficient alkene, MMA. Also, the preference of the MMA-derived radical for MMA is attenuated by a greater steric inhibition between the bulkier radical and bulkier alkene.

The r values show that an equimolar mixture of MMA and acrylonitrile will give a copolymer with more MMA units and further that most of the acrylonitrile units will be isolated between MMA units. In order to produce a copolymer with equal representation, the concentration of acrylonitrile would have to be higher in the monomer mixture.

Radical	Monomer	r
Ph (structure)	Ph (structure)	0.52
MeO, O (structure)	MeO, O (structure)	0.46

In this case, both radicals show a 2-fold preference for the opposite monomer. Here, recognizing the more stable radical is more difficult, since the styrene-derived radical is less electron-poor and more delocalized but only secondary, lacking the extra methyl of MMA. Therefore, we might best explain the fairly weak preference in both directions as arising from Coulombic effects: the more electron-rich radical prefers the more electron-poor alkene, and the more electron-poor radical prefers the more electron-rich alkene.

These r values show that these monomers will be incorporated with similar efficiency into the copolymer and that they will tend to alternate, though the alternation will not be complete.

Radical	Monomer	r
Ph (structure)	Ph (structure)	55
OAc (structure)	OAc (structure)	0.01

Clearly, styrene is a much more reactive monomer than vinyl acetate, being selected with 55-fold and 100-fold preferences. Though the oxygen lone pairs of the acetate can donate electron density to both the radical center and the alkene through resonance, the inductive electron withdrawing ability is the more important (as can be seen by inspecting various substituent constants for acetate). Therefore, the vinyl acetate-derived radical is less electron-rich and less stable, leading to the strong preference for the styrene monomer.

As copolymer is formed from these monomers, little vinyl acetate will be incorporated unless this monomer is present in a much greater concentration.

Radical	Monomer	r
		0.015
		20

As in the previous pairing, vinyl acetate is the less reactive monomer, though by a somewhat smaller margin. Comparison of substituent constant shows that acetate and carbomethoxy are very similar in both their electron-withdrawing and resonance stabilizing strengths. Therefore, the important difference in this case appears to be the extra methyl substituent in MMA, leading to the more stable radical. As noted in the first pairing, though MMA is preferred by both radicals, the smaller preference by the MMA radical can be attributed to a steric effect.

Unless the vinyl acetate is boosted in the monomer mixture, MMA units will dominate the copolymer.

11. Corannulene, even in its bowl-shaped conformation, has high symmetry, possessing 5 mirror planes and a C_5 axis. However, a single substituent is enough to destroy all symmetry.

A strategy for investigating the inversion barrier is to use a substituent that contains like subgroups that will be exchanged by the inversion process. Consider R = CH₂Cl. Based on our analyses of topicities in Chapter 6, we would expect the two hydrogens of this group to be diastereotopic. The molecule is asymmetric, and so no symmetry operation can interconvert the hydrogens. It may be less obvious in this system that no rotation about the C-C bond can interconvert them either. If we suppose that the center of the bowl is up and the rim down, we can see that the two H's are different – one is aimed toward the convex face and the other toward the concave face. It's true that a 180° rotation does change the orientations with respect to the bowl, but the conformation we get is different than the first one and the two H's are therefore not exchanged.

Inversion of the bowl would exchange the environments of the two H's. Therefore, taking NMR spectra of this compound at different temperatures should allow determination of the inversion barrier. At temperatures where the inversion is slow relative to the NMR time scale, the two H's should give different signals. At higher temperatures, where the inversion is faster, the two signals should coalesce into one averaged signal. Through line-shape analysis of these dynamic NMR data, the barrier could be determined.

12. The reason corannulene is bowl-shaped is that the five-membered ring leads to higher strain energy in the planar conformation. We can see why this happens by calculating bond angles for the fully symmetric, planar geometry. The interior angles in a regular pentagon are 108°. This means that the adjacent interior angles in the hexagons must be (360°-108°)/2 = 126°. These angles are clearly not optimum for an sp^2 center, and the remaining angles will necessarily also deviate from 120°.

An effect that favors the planar structure is the overlap of the p orbitals that leads to both π bonding and aromaticity. In the bowl-shaped geometry, the sp^2 centers are pyramidalized somewhat towards sp^3, such that the π bonding depends on the overlap of slightly hybridized and non-parallel orbitals. Therefore, this effect presumably results in a smaller barrier to inversion than would otherwise be observed.

13. From the text, we know that C_{60} contains 12 five-membered rings and that these rings are all isolated from each other (non-fused). Therefore, since each ring contains 5 bonds that must be classified as 6-5 bonds, there are in total 12 × 5 = 60 bonds of the 6-5 type. The 6-6 bonds are all of the same type: bonds that connect two of the five-membered rings. If we counted 5 of these for each of the five-membered rings, we would be counting each of these bonds twice. So dividing by 2 to avoid this double-counting, we get (12 × 5)/2 = 30 bonds of the 6-6 type.

14. Two effects may be important. First, the alkyl chains tilt from the surface normal to fill space. Even if the S atoms are close-packed on the surface, the alkyl chains have a smaller cross-section than S, leaving some space between the chains. This space is eliminated by tilting the chains, allowing the attractive van der Waals forces between the chains to be optimized.

Also worth considering are the bonding interactions between the S atoms and the surface metal atoms, though these effects are less clear. A simple analysis might say that the large Au–S–C bond angle shown in A or the bent surface interaction of B would be inferior to the more optimum arrangement of C, having a tilted alkyl chain. However, the surface bonding interactions are not so simple. For example, the interactions are not purely covalent and each S might interact with several Au centers. So the space-filling argument is more obviously important.

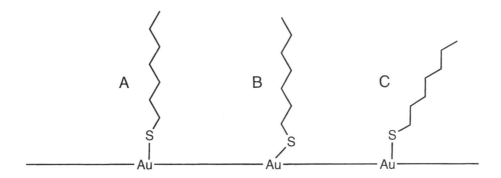

15. From largest to smallest: $M_w(B) > M_w(A) > M_n(B) = M_n(A)$. As for any distributions with finite widths, $M_w(A) > M_n(A)$ and $M_w(B) > M_n(B)$. The larger chains present in sample B will be weighted most heavily in $M_w(B)$, giving this a higher value than $M_w(A)$. M_n values are always at the center of a symmetric distribution, and since these two distributions are centered at the same place, $M_n(B) = M_n(A)$.

16. a. At some point in the polymerization, the chains that happen to be at the high end of the molecular weight distribution, due to their lower termination rate, are more likely to keep growing. Shorter chains are more likely to be terminated. This can lead to a bimodal distribution. (Depending on the relative rates, this situation might or might not lead to a bimodal distribution.)

    b. The heaviest chains are given the greatest weight in the calculation of M_w. Though the small peak at high molecular weight will affect both M_n and M_w, it will have a greater effect on M_w.

17. Increased branching in a polymer tends to prevent polymer chains from packing tightly and forming crystalline regions. With reduced interactions between chains, the polymer melts more readily. Cross-links are covalent links between chains, so instead of reducing interactions, they hold the chains together very strongly. Even a low cross-link density leads to a very marked increase in melting point. With a high enough cross-link density, the polymer might not melt at any temperature.

18. This molecule, though possessing an appropriate shape for liquid crystallinity, is quite rigid and lacks any flexible portions that would help to prevent crystallization. Flexible alkyl chains could be easily attached through the anhydride portions of the molecule:

19. a. The liquid crystals are aligned and oriented by the magnetic field of the NMR spectrometer.

b. The main characteristic is molecular shape. A rod-shape molecule will be more readily aligned than a more spherical one. Another characteristic that can be important is the presence of polar functional groups, whose dipoles will tend to cause alignment with respect to neighboring dipoles.

20. This relationship tells us that as the degree of polymerization gets larger the melting temperature increases. The linear relationship between the inverse melting temperature and the inverse degree of polymerization gives a slope proportional to $R/\Delta H_{fus}$. This tells us that as the energy of interaction between the monomer units in the polymer increases, the slope decreases and the degree of polymerization therefore has less of an effect on changing the melting temperature. This means that the polymer chain end groups have a decreased effect on the melting temperature as the interactions between neighboring monomer units increase. Further, the exercise tells us that this relationship only holds for high molecular weight polymers, meaning those with large monomers or a high DP to start. Taken together, this implies that the melting temperature is related to the energy of interaction between monomers within the chains, and the more end groups within the sample, the lower the melting temperature. The y-intercept gives the reciprocal melting temperature as DP goes to infinity (as 1/DP goes to zero), meaning a pure sample of the polymer with no end groups.

21. For the polymer to precipitate at higher temperatures, the entropy of solvation must be negative. Given that $\Delta G = \Delta H - T\Delta S$, the $T\Delta S$ term will begin to dominate at higher temperatures and will overwhelm any favorable enthalpy of solvation, causing the polymer to precipitate. This may be expected for a highly solvophobic polymer, because precipitation releases solvent molecules to bulk solution, making the precipitation entropically favored. This reasoning reinforces the notion that the solvation must be entropically disfavored, because solvation is the opposite of precipitation.

22. The slight deviation from Eq. 13.9 at the beginning of a step growth polymerization can be attributed to differences in rate constants. The rate constant for A-A condensing with B-B, which is the very initial step in the polymerization, can be slightly different than the rate constant for condensing P-A with B-P', where P and P' represent arbitrary lengths of polymer that are attached to A and B respectively. Once all A and B are attached to polymer stands, the rate constants for further reaction become so similar that Eq. 13.9 holds.

23. Here are three reasons: (1) Both termination and propagation are second-order processes: rate(prop.) = k_p[monomer][free radical] and rate(term.) = k_t[free radical]^2. The monomer concentration is high and relatively constant, but the free radical concentration can vary much more. Therefore, the termination rate, with its second-order dependence on free radical concentration, would be expected to vary more than propagation.

(2) The nature of the free radicals that are present changes dramatically through a polymerization reaction. Initially, they are small molecules or oligomers, but as polymerization proceeds, at least some of them become very large molecules.

(3) As polymerization proceeds, the viscosity changes dramatically. This has a very large effect on the rate that radicals diffuse towards each other and terminate. It has a much smaller effect on the propagation rate, since the monomer concentration remains high and diffusion is not as important – any given free radical is likely to be already next to a monomer molecule.

CHAPTER

14

Advanced Concepts in Electronic Structure Theory

SOLUTIONS TO EXERCISES

1. The procedure, outlined in the text, is filled out here. We start by plugging Eq. 14.38,

$$\psi = c_A\phi_A + c_B\phi_B$$

into Eq. 14.11,

$$E = \langle\psi|H|\psi\rangle/\langle\psi|\psi\rangle$$

which gives

$$E = \frac{\langle c_A\phi_A + c_B\phi_B|H|c_A\phi_A + c_B\phi_B\rangle}{\langle c_A\phi_A + c_B\phi_B|c_A\phi_A + c_B\phi_B\rangle}$$

Remembering the meaning of the braket notation:

$$\langle\psi|H|\psi\rangle = \int \psi * H\psi d\tau$$

guides us in correctly multiplying out the terms.

$$E = \frac{\langle c_A\phi_A|H|c_A\phi_A\rangle + \langle c_A\phi_A|H|c_B\phi_B\rangle + \langle c_B\phi_B|H|c_A\phi_A\rangle + \langle c_B\phi_B|H|c_B\phi_B\rangle}{\langle c_A\phi_A|c_A\phi_A\rangle + \langle c_A\phi_A|c_B\phi_B\rangle + \langle c_B\phi_B|c_A\phi_A\rangle + \langle c_B\phi_B|c_B\phi_B\rangle}$$

Taking the constants out of the integrals gives

$$E = \frac{c_A^2\langle\phi_A|H|\phi_A\rangle + c_Ac_B\langle\phi_A|H|\phi_B\rangle + c_Ac_B\langle\phi_B|H|\phi_A\rangle + c_B^2\langle\phi_B|H|\phi_B\rangle}{c_A^2\langle\phi_A|\phi_A\rangle + c_Ac_B\langle\phi_A|\phi_B\rangle + c_Ac_B\langle\phi_B|\phi_A\rangle + c_B^2\langle\phi_B|\phi_B\rangle}$$

Making the integral substitutions noted after Eq. 14.39, we get

$$E = \frac{c_A^2 H_{AA} + 2c_Ac_B H_{AB} + c_B^2 H_{BB}}{c_A^2 + 2c_Ac_B S + c_B^2} = \frac{N}{D}$$

Our goal now is to find the values of c_A and c_B that minimize the value of E. To do this, we take the partial derivatives of E with respect to c_A and c_B and set them equal to zero. Since E is in the form of a quotient (N/D), we use the quotient rule. Starting with c_A,

$$\frac{\partial E}{\partial c_A} = \frac{\partial}{\partial c_A}\left(\frac{N}{D}\right) = \frac{D\dfrac{\partial N}{\partial c_A} - N\dfrac{\partial D}{\partial c_A}}{D^2} = 0$$

$$D\frac{\partial N}{\partial c_A} - N\frac{\partial D}{\partial c_A} = 0$$

$$\frac{\partial N}{\partial c_A} - \frac{N}{D}\frac{\partial D}{\partial c_A} = 0$$

$$\frac{\partial N}{\partial c_A} - E\frac{\partial D}{\partial c_A} = 0$$

$$\frac{\partial\left(c_A^2 H_{AA} + 2c_Ac_B H_{AB} + c_B^2 H_{BB}\right)}{\partial c_A} - E\frac{\partial\left(c_A^2 + 2c_Ac_B S + c_B^2\right)}{\partial c_A} = 0$$

$$2c_A H_{AA} + 2c_B H_{AB} - E\left(2c_A + 2c_B S\right) = 0$$

$$c_A H_{AA} + c_B H_{AB} - c_A E - c_B SE = 0$$

$$c_A\left(H_{AA} - E\right) + c_B\left(H_{AB} - SE\right) = 0$$

Repeating this procedure for c_B gives

$$c_A\left(H_{AB} - SE\right) + c_B\left(H_{BB} - E\right) = 0$$

The last two equations are called the secular equations, and we can now represent them with a determinant equation.

$$\begin{vmatrix} H_{AA} - E & H_{AB} - SE \\ H_{AB} - SE & H_{BB} - E \end{vmatrix} = 0$$

At this point, we narrow the problem by considering only the symmetric two-orbital mixing problem, such that $H_{AA} = H_{BB}$.

$$\begin{vmatrix} H_{AA} - E & H_{AB} - SE \\ H_{AB} - SE & H_{AA} - E \end{vmatrix} = 0$$

We then solve the system of equations by reducing the determinant:

$$\left(H_{AA} - E \right)^2 - \left(H_{AB} - SE \right)^2 = 0$$

The roots of this equation are found by factoring:

$$\left[\left(H_{AA} - E \right) + \left(H_{AB} - SE \right) \right] \left[\left(H_{AA} - E \right) - \left(H_{AB} - SE \right) \right] = 0$$

$$\left(H_{AA} - E \right) + \left(H_{AB} - SE \right) = 0 \quad \text{or} \quad \left(H_{AA} - E \right) - \left(H_{AB} - SE \right) = 0$$

$$H_{AA} + H_{AB} - E \left(1 + S \right) = 0 \quad \text{or} \quad H_{AA} - H_{AB} - E \left(1 - S \right) = 0$$

So the energies of the two orbitals, ε_1 and ε_2, are

$$E = \frac{H_{AA} + H_{AB}}{1 + S} = \varepsilon_1 \quad \text{or} \quad E = \frac{H_{AA} - H_{AB}}{1 - S} = \varepsilon_2$$

To find c_A and c_B for the orbital with $E = \varepsilon_1$, we substitute into one of the secular equations. (It doesn't matter which one.)

$$c_A \left(H_{AA} - \frac{H_{AA} + H_{AB}}{1 + S} \right) + c_B \left(H_{AB} - S \frac{H_{AA} + H_{AB}}{1 + S} \right) = 0$$

$$c_A \left(\frac{H_{AA} + H_{AA}S - H_{AA} - H_{AB}}{1 + S} \right) + c_B \left(\frac{H_{AB} + H_{AB}S - H_{AA}S - H_{AB}S}{1 + S} \right) = 0$$

$$c_A \left(H_{AA}S - H_{AB} \right) + c_B \left(H_{AB} - H_{AA}S \right) = 0$$

$$c_A - c_B = 0$$

$$c_A = c_B$$

For $E = \varepsilon_2$,

$$c_A \left(H_{AA} - \frac{H_{AA} - H_{AB}}{1-S} \right) + c_B \left(H_{AB} - S \frac{H_{AA} - H_{AB}}{1-S} \right) = 0$$

$$c_A \left(\frac{H_{AA} - H_{AA}S - H_{AA} + H_{AB}}{1-S} \right) + c_B \left(\frac{H_{AB} - H_{AB}S - H_{AA}S + H_{AB}S}{1-S} \right) = 0$$

$$c_A (H_{AB} - H_{AA}S) + c_B (H_{AB} - H_{AA}S) = 0$$

$$c_A + c_B = 0$$

$$c_A = -c_B$$

So the two wavefunctions are

$$\psi_1 = c_A \phi_A + c_A \phi_B \quad \text{and} \quad \psi_2 = c_A \phi_A - c_A \phi_B$$

We can get the value of c_A by normalizing the wavefunctions. For ψ_1,

$$\langle c_A \phi_A + c_A \phi_B | c_A \phi_A + c_A \phi_B \rangle = 1$$

$$\langle c_A \phi_A | c_A \phi_A \rangle + \langle c_A \phi_A | c_A \phi_B \rangle + \langle c_A \phi_B | c_A \phi_A \rangle + \langle c_A \phi_B | c_A \phi_B \rangle = 1$$

$$c_A^2 \left(\langle \phi_A | \phi_A \rangle + \langle \phi_A | \phi_B \rangle + \langle \phi_B | \phi_A \rangle + \langle \phi_B | \phi_B \rangle \right) = 1$$

$$c_A^2 (1 + S + S + 1) = 1$$

$$c_A = \frac{1}{\sqrt{2 + 2S}}$$

For ψ_2,

$$\langle c_A \phi_A - c_A \phi_B | c_A \phi_A - c_A \phi_B \rangle = 1$$

$$\langle c_A \phi_A | c_A \phi_A \rangle - \langle c_A \phi_A | c_A \phi_B \rangle - \langle c_A \phi_B | c_A \phi_A \rangle + \langle c_A \phi_B | c_A \phi_B \rangle = 1$$

$$c_A^2 \left(\langle \phi_A | \phi_A \rangle - \langle \phi_A | \phi_B \rangle - \langle \phi_B | \phi_A \rangle + \langle \phi_B | \phi_B \rangle \right) = 1$$

$$c_A^2 (1 - S - S + 1) = 1$$

$$c_A = \frac{1}{\sqrt{2 - 2S}}$$

The final wavefunctions and energies for the two-orbital mixing problem are

$$\psi_1 = \frac{1}{\sqrt{2+2S}}\left(\phi_A + \phi_B\right) \text{ with } \varepsilon_1 = \frac{H_{AA} + H_{AB}}{1+S} \text{ and}$$

$$\psi_2 = \frac{1}{\sqrt{2-2S}}\left(\phi_A - \phi_B\right) \text{ with } \varepsilon_2 = \frac{H_{AA} - H_{AB}}{1-S}$$

2. a. As the number of lobes separated by nodal surfaces increases, the energy increases. The lowest energy orbital has only one lobe and no nodes: E. Going up from there is A, C, and F, each with two lobes, D and G, each with three lobes, and B, with four lobes. Thus, orbital B is the HOMO, having significant p character on O. This orbital may be considered to represent an O lone pair, with some C–H bonding character as well. Electrophiles or Lewis acids, which will tend to interact most strongly with the HOMO, will therefore tend to approach this O lone pair (just as we know they should).

b. The LUMO, though having complex nodal surfaces and small interior lobes, does have a large lobe surrounding the H attached to O. Therefore, nucleophiles or Lewis bases, which will tend to interact most strongly with the LUMO, will tend to approach this H atom, resulting in H-bonding interactions and perhaps ultimately even deprotonation.

3. Though we usually think of methane as a small molecule, it has quite a few particles that need to be considered in the full Hamiltonian: 5 nuclei and 10 electrons. Our job is simplified by symmetry in the positions of nuclei, allowing us to combine like interactions in common terms. For example, even though there are 4 C–H bonds, r_{CH} can be taken to have a single value.

$$H = \left(\text{kin. E of nuc.}\right) + \left(\text{nuc. - nuc. rep.}\right) + \left(\text{kin. E of } e^-\right) + \left(\text{nuc. - } e^- \text{ attractions}\right) + \left(e^- - e^- \text{ rep.}\right)$$

For convenience, we will separate the C from the H's in the first and fourth terms. Also, we will separate C–H and H–H interactions in the second term.

$$H = \left(-\frac{\hbar^2}{2}\frac{\nabla_C^2}{M_C} - \sum_{A=1}^{4}\frac{\hbar^2}{2}\frac{\nabla_A^2}{M_A}\right) + \left(\sum_{A=1}^{4}\frac{e^2 Z_C Z_H}{r_{CH}} + \sum_{A<B}\frac{e^2 Z_H Z_H}{r_{HH}}\right) + \left(-\frac{\hbar^2}{2m}\sum_{i=1}^{10}\nabla_i^2\right)$$

$$+ \left(-\sum_{i=1}^{10}\frac{e^2 Z_C}{r_{Ci}} - \sum_{A=1}^{4}\sum_{i=1}^{10}\frac{e^2 Z_H}{r_{Ai}}\right) + \left(\sum_{i<j}\frac{e^2}{r_{ij}}\right)$$

$$H = -\frac{\hbar^2}{2}\left(\frac{\nabla_C^2}{M_C} + \frac{4\nabla_H^2}{M_H}\right) + e^2\left(\frac{4 Z_C Z_H}{r_{CH}} + \frac{6 Z_H Z_H}{r_{HH}}\right) - \frac{\hbar^2}{2m}\sum_{i=1}^{10}\nabla_i^2 - \sum_{i=1}^{10}\frac{e^2 Z_C}{r_{Ci}} - \sum_{A=1}^{4}\sum_{i=1}^{10}\frac{e^2 Z_H}{r_{Ai}} + \sum_{i<j}\frac{e^2}{r_{ij}}$$

Further simplification is done by recognizing that $Z_C = 6$ and $Z_H = 1$.

$$H = -\frac{\hbar^2}{2}\left(\frac{\nabla_C^2}{M_C} + \frac{4\nabla_H^2}{M_H}\right) + e^2\left(\frac{24}{r_{CH}} + \frac{6}{r_{HH}}\right) - \frac{\hbar^2}{2m}\sum_{i=1}^{10}\nabla_i^2 - \sum_{i=1}^{10}\frac{6e^2}{r_{Ci}} - \sum_{A=1}^{4}\sum_{i=1}^{10}\frac{e^2}{r_{Ai}} + \sum_{i<j}\frac{e^2}{r_{ij}}$$

It is interesting to note that the summation in the last term, the electron-electron repulsion term, includes 45 subterms! (There are 45 pairwise interactions of 10 electrons.)

4. The Morse potential, which reflects the total energy, is just the sum of the potential (V_g) and kinetic (T_g) energy components. By lowering either component, the sum is also lowered, resulting in a stronger bond (higher BDE).

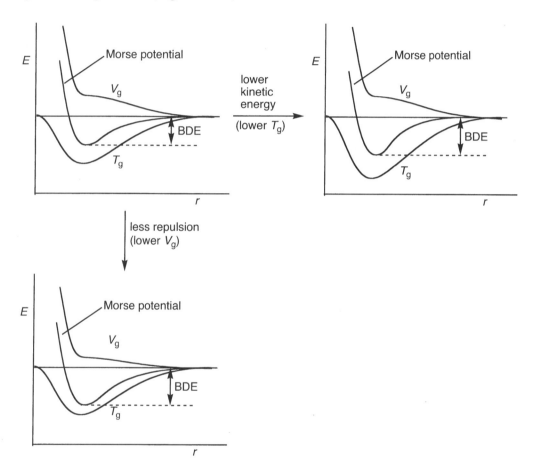

5. In an He atom, the lone valence orbital (1s) is completely filled. Therefore, any interaction between two He atoms necessarily involves a filled-filled (2c, 4e⁻) interaction. These interactions are always repulsive, since the antibonding MO is always destabilized more than the bonding MO is stabilized.

Another way to look at this is in terms of potential and kinetic energy. Since the repulsive electrostatic interactions always outweigh the attractive ones, in order for a bond to form the kinetic energy of the electrons must overcompensate. In He$_2$, two electrons experience a kinetic energy stabilization, but the other two are less stable than they are in the isolated atoms. Not only does the kinetic energy fail to compensate for the increased potential energy, but it is increased as well!

6. "Second-order Moller-Plesset" refers to an MO method of calculation. Specifically, MP2 is a perturbation method designed to provide electron correlation, which is not included in the HF method.

"Hybrid density functional theory" is a calculation method that is different from MO theory, though the term "hybrid" indicates that some aspects of the HF method are retained. DFT is a method based on electron density rather than orbital wavefunctions.

"Basis sets" are collections of functions intended to approximate atomic orbitals, such that a calculation method can use the LCAO approximation to derive accurate MOs from the available basis functions.

"6-31G*" is a particular basis set. Each number and character in this name conveys a meaning with respect to how the basis set is constructed. Beyond this, the name conveys exactly the selection of basis functions that is used, such that anyone else using the same method and basis set should obtain the same results.

"6-311G*" is another particular basis set. The similar name shows that it is related to 6-31G*, and the extra number shows that this one is more advanced in some respect (contains more basis functions). In particular, this is a 3-fold split valence basis set as opposed to 6-31G*, which is 2-fold (not including the polarization functions).

"AM1" and "PM3" are successive improvements of the MNDO semiempirical method (Austin Method 1 and Parametric Method 3).

"SCF" stands for self-consistent field and refers to an iterative method for MO calculations involving alternate operations of changing the MOs (the coefficients of the various basis functions) and then calculating the energy integrals. In this manner, the orbitals can be optimized to give the lowest total energy.

7. For allyl ($N = 3$),

$$\varepsilon_1 = \alpha + 2\beta \cos\left(\frac{\pi}{4}\right) = \alpha + 2\beta(0.7071) = \alpha + 1.414\beta$$

$$\varepsilon_2 = \alpha + 2\beta \cos\left(\frac{2\pi}{4}\right) = \alpha + 2\beta(0) = \alpha$$

$$\varepsilon_3 = \alpha + 2\beta \cos\left(\frac{3\pi}{4}\right) = \alpha + 2\beta(-0.7071) = \alpha - 1.414\beta$$

For butadiene ($N = 4$),

$$\varepsilon_1 = \alpha + 2\beta \cos\left(\frac{\pi}{5}\right) = \alpha + 2\beta(0.8090) = \alpha + 1.618\beta$$

$$\varepsilon_2 = \alpha + 2\beta\cos\left(\frac{2\pi}{5}\right) = \alpha + 2\beta(0.3090) = \alpha + 0.618\beta$$

$$\varepsilon_3 = \alpha + 2\beta\cos\left(\frac{3\pi}{5}\right) = \alpha + 2\beta(-0.3090) = \alpha - 0.618\beta$$

$$\varepsilon_4 = \alpha + 2\beta\cos\left(\frac{4\pi}{5}\right) = \alpha + 2\beta(-0.8090) = \alpha - 1.618\beta$$

8. *Allyl.* The easiest way to set up any secular determinant is using $-x$, 1, and 0.

$$\begin{vmatrix} -x & 1 & 0 \\ 1 & -x & 1 \\ 0 & 1 & -x \end{vmatrix} = 0$$

We evaluate the determinant by cofactor expansion along the first row.

$$-x\begin{vmatrix} -x & 1 \\ 1 & -x \end{vmatrix} - \begin{vmatrix} 1 & 1 \\ 0 & -x \end{vmatrix} = 0$$

$$-x(x^2 - 1) - (-x) = 0$$

$$-x^3 + 2x = 0$$

$$x = 0 \text{ or } x^2 = 2$$

$$x = 0 \text{ or } \pm\sqrt{2}$$

$$\frac{-(\alpha - E)}{\beta} = 0 \text{ or } \frac{-(\alpha - E)}{\beta} = \pm\sqrt{2}$$

$$E = \alpha \text{ or } E = \alpha \pm \sqrt{2}\beta$$

To get the coefficients for each MO, we substitute the values of x into the secular equations, corresponding to rows of the initial determinant. For example, the first row corresponds to the equation:

$$(-x)c_1 + c_2 = 0$$

For $x = 0$, $c_2 = 0$. Then using the second secular equation:

$$c_1 + (-x)c_2 + c_3 = 0$$

$$c_1 = -c_3$$

Normalizing: $c_1^2 + c_2^2 + c_3^2 = 1$

$$c_1^2 + (0)^2 + (-c_1)^2 = 1$$

So $c_1 = \dfrac{1}{\sqrt{2}}$, $c_2 = 0$, $c_3 = -\dfrac{1}{\sqrt{2}}$ for $E = \alpha$

(An equivalent set of coefficients would have $c_1 = -\dfrac{1}{\sqrt{2}}$ and $c_3 = \dfrac{1}{\sqrt{2}}$, since the signs of the wavefunctions are arbitrary except in a relative sense.)

For $x = \sqrt{2}$ (*i.e.*, $E = \alpha + \sqrt{2}\beta$), the first secular equation gives

$$-\sqrt{2}c_1 + c_2 = 0 \text{, so } c_2 = \sqrt{2}c_1$$

The third secular equation, $c_2 + (-x)c_3 = 0$, gives

$$c_2 - \sqrt{2}c_3 = 0 \text{, so } c_2 = \sqrt{2}c_3 \text{ and therefore } c_1 = c_3$$

Normalizing: $c_1^2 + \left(\sqrt{2}c_1\right)^2 + c_1^2 = 1$

$$4c_1^2 = 1 \text{, so } c_1 = \frac{1}{2}, \ c_2 = \frac{\sqrt{2}}{2} = \frac{1}{\sqrt{2}}, \ c_3 = \frac{1}{2} \text{ for } E = \alpha + \sqrt{2}\beta$$

For $x = -\sqrt{2}$ (*i.e.*, $E = \alpha - \sqrt{2}\beta$), the first secular equation gives

$$\sqrt{2}c_1 + c_2 = 0 \text{, so } c_2 = -\sqrt{2}c_1$$

The third secular equation gives

$$c_2 + \sqrt{2}c_3 = 0 \text{, so } c_2 = -\sqrt{2}c_3 \text{ and therefore } c_1 = c_3$$

Normalizing: $c_1^2 + \left(-\sqrt{2}c_1\right)^2 + c_1^2 = 1$

$$c_1 = \frac{1}{2}, \ c_2 = -\frac{\sqrt{2}}{2} = -\frac{1}{\sqrt{2}}, \ c_3 = \frac{1}{2} \text{ for } E = \alpha - \sqrt{2}\beta$$

All energies and coefficients agree with Figure 14.15.

Butadiene.

$$\begin{vmatrix} -x & 1 & 0 & 0 \\ 1 & -x & 1 & 0 \\ 0 & 1 & -x & 1 \\ 0 & 0 & 1 & -x \end{vmatrix} = 0$$

Cofactor expansion along the first row gives

$$-x \begin{vmatrix} -x & 1 & 0 \\ 1 & -x & 1 \\ 0 & 1 & -x \end{vmatrix} - \begin{vmatrix} 1 & 1 & 0 \\ 0 & -x & 1 \\ 0 & 1 & -x \end{vmatrix} = 0$$

$$-x \left(-x \begin{vmatrix} -x & 1 \\ 1 & -x \end{vmatrix} - \begin{vmatrix} 1 & 1 \\ 0 & -x \end{vmatrix} \right) - \begin{vmatrix} -x & 1 \\ 1 & -x \end{vmatrix} = 0$$

$$-x \left(-x \left(x^2 - 1 \right) - (-x) \right) - \left(x^2 - 1 \right) = 0$$

$$-x \left(-x^3 + 2x \right) - x^2 + 1 = 0$$

$$x^4 - 3x^2 + 1 = 0$$

We can solve this equation by setting $y = x^2$ and then using the quadratic formula:

$$y^2 - 3y + 1 = 0$$

$$y = \frac{-(-3) \pm \sqrt{(-3)^2 - 4(1)(1)}}{2(1)} = \frac{3 \pm \sqrt{5}}{2} = 2.618 \text{ or } 0.382$$

So $x = \pm\sqrt{2.618}$ or $\pm\sqrt{0.382}$

$x = \pm 1.618$ or ± 0.618

The secular equations are

$$(-x)c_1 + c_2 = 0, \ c_1 + (-x)c_2 + c_3 = 0, \ c_2 + (-x)c_3 + c_4 = 0, \text{ and } c_3 + (-x)c_4 = 0$$

For $x = 1.618$, the first and fourth equations give

$$c_2 = 1.618c_1 \text{ and } c_3 = 1.618c_4$$

Substituting for both x and c_3 in the third equation gives

$$c_2 - 1.618(1.618c_4) + c_4 = 0$$

$c_2 = 1.618c_4$, so $c_2 = c_3$ and $c_1 = c_4$

Normalizing: $c_1^2 + c_2^2 + c_3^2 + c_4^2 = c_1^2 + (1.618c_1)^2 + (1.618c_1)^2 + c_1^2 = 1$

$7.236c_1^2 = 1$, so $c_1 = 0.37$ (We could equivalently use $c_1 = -0.37$, but this way the signs match Figure 14.15.)

So for $x = 1.618$, $E = \alpha + 1.618\beta$, with $c_1 = c_4 = 0.37$ and $c_2 = c_3 = (1.618)0.37 = 0.60$

For $x = -1.618$, the first and fourth equations give

$$c_2 = -1.618c_1 \text{ and } c_3 = -1.618c_4$$

Substituting for both x and c_3 in the third equation gives

$$c_2 + 1.618(-1.618c_4) + c_4 = 0$$

$c_2 = 1.618c_4$, so $c_2 = -c_3$ and $c_1 = -c_4$

Normalizing: $c_1^2 + (-1.618c_1)^2 + (1.618c_1)^2 + (-c_1)^2 = 1$

$7.236c_1^2 = 1$, so $c_1 = -0.37$ (matching the sign in Figure 14.15)

So for $x = -1.618$, $E = \alpha - 1.618\beta$, with $c_1 = -0.37$, $c_2 = 0.60$, $c_3 = -0.60$, and $c_4 = 0.37$

For $x = 0.618$, the first and fourth equations give

$$c_2 = 0.618c_1 \text{ and } c_3 = 0.618c_4$$

Substituting for both x and c_3 in the third equation gives

$$c_2 - 0.618(0.618c_4) + c_4 = 0$$

$c_2 = -0.618c_4$, so $c_2 = -c_3$ and $c_1 = -c_4$

Normalizing: $c_1^2 + (0.618c_1)^2 + (-0.618c_1)^2 + (-c_1)^2 = 1$

$2.764c_1^2 = 1$, so $c_1 = -0.60$ (matching the sign in Figure 14.15)

So for $x = 0.618$, $E = \alpha + 0.618\beta$, with $c_1 = -0.60$, $c_2 = -0.37$, $c_3 = 0.37$, and $c_4 = 0.60$

For $x = -0.618$, the first and fourth equations give

$c_2 = -0.618c_1$ and $c_3 = -0.618c_4$

Substituting for both x and c_3 in the third equation gives

$c_2 + 0.618(-0.618c_4) + c_4 = 0$

$c_2 = -0.618c_4$, so $c_2 = c_3$ and $c_1 = c_4$

Normalizing: $c_1^2 + (-0.618c_1)^2 + (-0.618c_1)^2 + c_1^2 = 1$

$2.764c_1^2 = 1$, so $c_1 = -0.60$ (matching the sign in Figure 14.15)

So for $x = -0.618$, $E = \alpha - 0.618\beta$, with $c_1 = c_4 = -0.60$ and $c_2 = c_3 = 0.37$

The energies and coefficients match completely with Figure 14.15 for all four MOs.

9. Pentadienyl is an odd alternant hydrocarbon, so it will have an NBMO ($E = \alpha$) that has coefficients only on the starred atoms:

The other four π MOs will be paired with energies $E = \alpha \pm x\beta$ and differing within the pairs only by the sign of the coefficients for the starred atoms. Therefore, the lowest energy MO, which has no nodes, is paired with the highest energy MO, which has four nodes:

Finally, the second lowest MO, having one node, is paired with the second highest MO, having three nodes:

10. For an odd AH, the coefficients of the NBMO can be found quite easily. First, we label such that we have more starred than non-starred atoms – the coefficients will only be non-zero on the starred atoms. These coefficients adjacent to each non-starred atom must sum to zero.

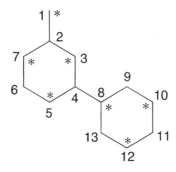

The easiest place to start is with non-starred atoms that are adjacent to only two starred atoms. For these atoms the adjacent coefficients must be equal and opposite in sign. For example, if $c_8 = a$, then $c_{10} = -a$, so that the sum of coefficients around atom 9 sum to zero. Summing at atom 11 then forces $c_{12} = a$. But then the sum at atom 13 is $c_8 + c_{12} = 2a = 0$. So $a = 0$, and there is no NBMO character on the phenyl substituent! (If we had drawn resonance structures before we started, we would have expected this result, since there is no way to draw a resonance structure with the radical in the phenyl ring.)

Since $c_8 = 0$, this NBMO should end up being the same as for benzyl radical. Let's say $c_5 = b$, then both c_3 (summing at atom 4) and c_7 (summing at atom 6) must be equal to $-b$. Summing at atom 2 gives $c_1 = 2b$. Normalizing gives

$$(2b)^2 + 2(-b)^2 + b^2 = 1$$

$$7b^2 = 1$$

$$b = \frac{1}{\sqrt{7}}$$

$$\Psi_{NBMO} = \frac{1}{\sqrt{7}}\left(2\phi_1 - \phi_3 + \phi_5 - \phi_7\right)$$

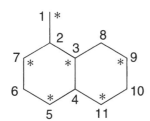

Let's start by saying $c_9 = a$. Then $c_3 = c_{11} = -a$ (summing at atoms 8 and 10). Summing at atom 4 gives $c_5 = 2a$. Then $c_7 = -2a$ (summing at atom 6). Finally, summing at atom 2 gives $c_1 = 3a$. Normalization:

$$(3a)^2 + (-2a)^2 + (2a)^2 + 2(-a)^2 + a^2 = 1$$

$$20a^2 = 1$$

$$a = \frac{1}{\sqrt{20}} = \frac{1}{2\sqrt{5}}$$

$$\Psi_{NBMO} = \frac{1}{2\sqrt{5}}\left(3\phi_1 - \phi_3 + 2\phi_5 - 2\phi_7 + \phi_9 - \phi_{11}\right)$$

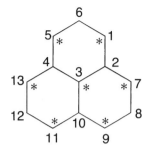

If we say $c_1 = a$, then summing at atom 6 gives $c_5 = -a$. Since this MO is not part of a degenerate set, it must reflect the symmetry of the molecule. Therefore, it must be symmetric with respect to the C_3 axis. (It is impossible for any orbital to be antisymmetric with respect to an odd axis. Try it!) This means that $c_1 = c_9 = c_{13} = a$ and $c_5 = c_7 = c_{11} = -a$. Summing at atom 2 (or 4 or 10) gives $c_3 = 0$. We can also tell that $c_3 = 0$ by noting that the orbital, as defined so far, is antisymmetric with respect to the vertical mirror plane that passes through atoms 6, 3, and 10. Normalization:

$$3(-a)^2 + 3a^2 = 1$$

$$6a^2 = 1$$

$$a = \frac{1}{\sqrt{6}}$$

$$\Psi_{NBMO} = \frac{1}{\sqrt{6}}\left(\phi_1 - \phi_5 - \phi_7 + \phi_9 - \phi_{11} + \phi_{13}\right)$$

You might wonder why there is no contribution from atom 3 when two perfectly good resonance structures can be drawn with the radical on the central C. We can rationalize this by noting that the aromaticity of all three rings is lost in these forms.

We start by saying $c_{11} = a$. It then follows that $c_9 = c_{13} = -a$ (summing at atoms 10 and 12) and $c_5 = 2a$ (summing at atom 8). Likewise, $c_3 = c_7 = -2a$ and $c_1 = 4a$. Normalization:

$$(4a)^2 + 2(-2a)^2 + (2a)^2 + 2(-a)^2 + a^2 = 1$$

$$31a^2 = 1$$

$$a = \frac{1}{\sqrt{31}}$$

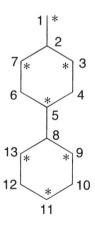

$$\Psi_{NBMO} = \frac{1}{\sqrt{31}}\left(4\phi_1 - 2\phi_3 + 2\phi_5 - 2\phi_7 - \phi_9 + \phi_{11} - \phi_{13}\right)$$

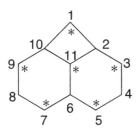

Again, it is easiest to start away from the *exo* methylene group. If we set $c_9 = a$, then $c_7 = c_{11} = -a$ (summing at atoms 8 and 10). Continuing along the bottom and summing at atoms 12 and 14 gives $c_{13} = a$ and $c_{15} = -a$. Summing at atoms 6, 4, and 2 gives, respectively, $c_5 = 2a$, $c_3 = -3a$, and $c_1 = 4a$. Normalization:

$$(4a)^2 + (-3a)^2 + (2a)^2 + 3(-a)^2 + 2(a)^2 = 1$$

$$34a^2 = 1$$

$$a = \frac{1}{\sqrt{34}}$$

$$\Psi_{\mathrm{NBMO}} = \frac{1}{\sqrt{34}}\left(4\phi_1 - 3\phi_3 + 2\phi_5 - \phi_7 + \phi_9 - \phi_{11} + \phi_{13} - \phi_{15}\right)$$

Starting with $c_5 = a$, summing at atom 4 gives $c_3 = -a$. We can get c_7 and c_9 by symmetry, but the NBMO could be either symmetric or antisymmetric with respect to the vertical mirror plane. We can see, however, that an antisymmetric orbital will not work. In this case, c_1 and c_{11} must both be zero. This would force c_3 also to be zero (summing at atom 2), as well as all other coefficients. Since the NBMO must be symmetric with respect to the plane, $c_7 = a$ and $c_9 = -a$. Summing at atom 6 gives $c_{11} = -2a$. Then summing at either atom 2 or atom 10 gives $c_1 = 3a$. Normalization:

$$(3a)^2 + 2(-a)^2 + 2(a)^2 + (-2a)^2 = 1$$

$$17a^2 = 1$$

$$a = \frac{1}{\sqrt{17}}$$

$$\Psi_{\mathrm{NBMO}} = \frac{1}{\sqrt{17}}\left(3\phi_1 - \phi_3 + \phi_5 + \phi_7 - \phi_9 - 2\phi_{11}\right)$$

11. The secular determinant for cyclobutadiene is set up and then solved:

$$\begin{vmatrix} -x & 1 & 0 & 1 \\ 1 & -x & 1 & 0 \\ 0 & 1 & -x & 1 \\ 1 & 0 & 1 & -x \end{vmatrix} = 0$$

$$-x\begin{vmatrix} -x & 1 & 0 \\ 1 & -x & 1 \\ 0 & 1 & -x \end{vmatrix} - \begin{vmatrix} 1 & 1 & 0 \\ 0 & -x & 1 \\ 1 & 1 & -x \end{vmatrix} - \begin{vmatrix} 1 & -x & 1 \\ 0 & 1 & -x \\ 1 & 0 & 1 \end{vmatrix} = 0$$

$$-x\left(-x\begin{vmatrix} -x & 1 \\ 1 & -x \end{vmatrix} - \begin{vmatrix} 1 & 1 \\ 0 & -x \end{vmatrix}\right) - \left(\begin{vmatrix} -x & 1 \\ 1 & -x \end{vmatrix} - \begin{vmatrix} 0 & 1 \\ 1 & -x \end{vmatrix}\right) - \left(\begin{vmatrix} 1 & -x \\ 0 & 1 \end{vmatrix} + \begin{vmatrix} -x & 1 \\ 1 & -x \end{vmatrix}\right) = 0$$

$$-x\left(-x\left(x^2 - 1\right) - (-x)\right) - \left(\left(x^2 - 1\right) - (-1)\right) - \left(1 + \left(x^2 - 1\right)\right) = 0$$

$$-x\left(-x^3 + 2x\right) - \left(x^2\right) - \left(x^2\right) = 0$$

$$x^4 - 4x^2 = 0$$

$$x^2 = 0 \text{ or } x^2 = 4$$

So the four roots are $x = 0$, $x = 0$, $x = 2$, and $x = -2$.

The four orbital energies are $E = \alpha$, $E = \alpha$, $E = \alpha + 2\beta$, and $E = \alpha - 2\beta$.

The secular equations are

$$(-x)c_1 + c_2 + c_4 = 0,\ c_1 + (-x)c_2 + c_3 = 0,\ c_2 + (-x)c_3 + c_4 = 0,\ \text{and}\ c_1 + c_3 + (-x)c_4 = 0$$

For $x = 0$, $c_2 = -c_4$ and $c_1 = -c_3$. The secular equations give no more restrictions than these, so we have some freedom in how we define these two degenerate, non-bonding MOs. The easiest is to exclude any bonding or antibonding interactions by alternately zeroing the two sets of coefficients. So for one MO, $c_2 = -c_4 = 0$, and for the other, $c_1 = -c_3 = 0$. Normalizing for the first MO,

$$c_1^2 + c_2^2 + c_3^2 + c_4^2 = c_1^2 + (-c_1)^2 = 1$$

$$c_1 = \frac{1}{\sqrt{2}} \text{ and } c_3 = -\frac{1}{\sqrt{2}} \text{ (while } c_2 = -c_4 = 0\text{)}$$

For the other $x = 0$ MO, $c_2 = \dfrac{1}{\sqrt{2}}$ and $c_4 = -\dfrac{1}{\sqrt{2}}$ (while $c_1 = -c_3 = 0$)

For the non-degenerate $x = 2$ and $x = -2$ MOs we hardly need the secular equations, since the fourfold symmetry of cyclobutadiene requires the four coefficients to have the same magnitude. We can therefore anticipate that the lower energy $x = 2$ MO has $c_1 = c_2 = c_3 = c_4$, with the higher energy $x = -2$ MO has $c_1 = -c_2 = c_3 = -c_4$. (Indeed, we can also obtain this result by adding or subtracting the secular equations.) In either case, the normalization is the same:

$$c_1^{\,2} + c_2^{\,2} + c_3^{\,2} + c_4^{\,2} = 4c_1^{\,2} = 1$$

So for $x = 2$, $c_1 = c_2 = c_3 = c_4 = \dfrac{1}{2}$, while for $x = -2$, $c_1 = c_3 = \dfrac{1}{2}$ and $c_2 = c_4 = -\dfrac{1}{2}$.

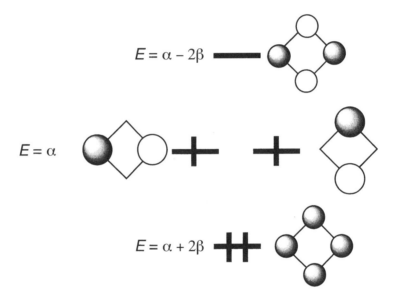

12. The PES shows four resolved ionizations from the valence shell. That result is reasonably consistent with the MO diagram in Figure 1.12, which shows electrons in five valence energy levels. It is inconsistent with six equivalent C–H bonds, since only two ionizations would be expected: one from C–H and one from C–C.

13. We must first do some electron counting:

HFe(PR$_3$)$_4$			CpFe(CO)$_2$	
Fe	8		Fe	8
H	1		Cp	5
PR$_3$	4 x 2 = 8		CO	2 x 2 = 4
total	17		total	17

Both $HFe(PR_3)_4$ and $CpFe(CO)_2$ can be considered as octahedral complexes with one open coordination site. This is less obvious for $CpFe(CO)_2$, with three coordination sites (d^2sp^3 hybrids) used by Cp and two used by CO ligands. In each case, this leaves one d^2sp^3 hybrid open, similar in shape to the open sp^3 hybrid in methyl, and containing one electron. For each complex, 10 electrons are used to form 5 bonds and 6 electrons reside in the t_{2g} set, leaving one electron for the open lobe.

14. The secular determinant for a three orbital mixing is:

$$\begin{vmatrix} H_{AA} - E & H_{AB} - SE & H_{AC} - SE \\ H_{BA} - SE & H_{BB} - E & H_{BC} - SE \\ H_{CA} - SE & H_{CB} - SE & H_{CC} - E \end{vmatrix} = 0$$

Due to symmetry, $H_{AA} = H_{BB} = H_{CC}$ and $H_{AB} = H_{BC} = H_{AC} = H_{BA} = H_{CB} = H_{CA}$. We are also neglecting overlap, so $S = 0$.

$$\begin{vmatrix} H_{AA} - E & H_{AB} & H_{AB} \\ H_{AB} & H_{AA} - E & H_{AB} \\ H_{AB} & H_{AB} & H_{AA} - E \end{vmatrix} = 0$$

For convenience, let $x = H_{AA} - E$ and $y = H_{AB}$

$$\begin{vmatrix} x & y & y \\ y & x & y \\ y & y & x \end{vmatrix} = 0$$

$$x \begin{vmatrix} x & y \\ y & x \end{vmatrix} - y \begin{vmatrix} y & y \\ y & x \end{vmatrix} + y \begin{vmatrix} y & x \\ y & y \end{vmatrix} = 0$$

$$x\left(x^2 - y^2\right) - y\left(xy - y^2\right) + y\left(y^2 - xy\right) = 0$$

$$\left(x^3 - xy^2\right) - \left(xy^2 - y^3\right) + \left(y^3 - xy^2\right) = 0$$

$$x^3 - 3xy^2 + 2y^3 = 0$$

$$\left(H_{AA} - E\right)^3 - 3\left(H_{AA} - E\right)H_{AB}^2 + 2H_{AB}^3 = 0$$

We can simplify this equation by realizing that $E = H_{AA} + zH_{AB}$.

$$\left(-zH_{AB}\right)^3 - 3\left(-zH_{AB}\right)H_{AB}^2 + 2H_{AB}^3 = 0$$

$$-z^3 H_{AB}{}^3 + 3z H_{AB}{}^3 + 2 H_{AB}{}^3 = 0$$

$$z^3 - 3z - 2 = 0$$

We can factor this equation by inspection. By trying small integers for z, we can quickly find, for example, that $z = 2$ is a root, giving us a start to factor the equation. (If required, there are direct methods for solving cubic equations. A web search on "cubic formula" turns up many resources that describe the straightforward but non-trivial methods. If all you want are the roots and are not interested in the process, search on "cubic equation calculator".)

$$(z - 2)(z^2 + 2z + 1) = 0$$

$$(z - 2)(z + 1)(z + 1) = 0$$

$$z = 2 \quad \text{or} \quad z = -1 \quad \text{or} \quad z = -1$$

So one of the orbitals has $E = H_{AA} + 2H_{AB}$, while the other two have $E = H_{AA} - H_{AB}$.

To get the orbital coefficients, the energies are substituted into the secular equations. For $E = H_{AA} + 2H_{AB}$, the first equation gives

$$c_1(H_{AA} - E) + c_2(H_{AB}) + c_3(H_{AB}) = 0$$

$$c_1(H_{AA} - H_{AA} - 2H_{AB}) + c_2(H_{AB}) + c_3(H_{AB}) = 0$$

$$2c_1 = c_2 + c_3$$

The second secular equation gives

$$c_1(H_{AB}) + c_2(H_{AA} - E) + c_3(H_{AB}) = 0$$

$$c_1(H_{AB}) + c_2(H_{AA} - H_{AA} - 2H_{AB}) + c_3(H_{AB}) = 0$$

$$2c_2 = c_1 + c_3$$

Plugging in c_2 to the first result gives

$$2c_1 = \left(\frac{c_1 + c_3}{2}\right) + c_3$$

$$c_1 = c_3$$

Substituting this result once more gives $c_1 = c_2 = c_3$. Normalizing, we get

$$c_1^2 + c_2^2 + c_3^2 = 1$$

$$3c_1^2 = 1$$

$$c_1 = \frac{1}{\sqrt{3}}$$

For $E = H_{AA} - H_{AB}$ (ψ_2 and ψ_3), the first secular equation gives

$$c_1(H_{AA} - H_{AA} + H_{AB}) + c_2(H_{AB}) + c_3(H_{AB}) = 0$$

$$c_1 + c_2 + c_3 = 0$$

The second and third equation gives the same result, such that no further information can be obtained from the secular equations. This situation is commonly encountered for degenerate MOs: no unique set of coefficients exist. The general solution is to make an assumption about one of the MOs, such that a consistent set of coefficients can then be obtained. In the present case, the MOs in question should clearly have a nodal plane, so let's assume that for ψ_2, atom 1 lies on the nodal plane, so that $c_1 = 0$. This gives $c_2 = -c_3$. Normalizing, we get

$$c_2^2 + c_3^2 = 1$$

$$2c_2^2 = 1$$

$$c_2 = \frac{1}{\sqrt{2}} \text{ and } c_3 = -\frac{1}{\sqrt{2}}$$

To obtain the coefficients of ψ_3, we can now use the fact that all MOs must be orthogonal.

$$\langle \psi_2 | \psi_3 \rangle = 0$$

$$\left\langle \frac{1}{\sqrt{2}}(\phi_2 - \phi_3) \middle| c_1\phi_1 + c_2\phi_2 + c_3\phi_3 \right\rangle = 0$$

$$\frac{1}{\sqrt{2}}\left(c_1\langle\phi_2|\phi_1\rangle + c_2\langle\phi_2|\phi_2\rangle + c_3\langle\phi_2|\phi_3\rangle - c_1\langle\phi_3|\phi_1\rangle - c_2\langle\phi_3|\phi_2\rangle - c_3\langle\phi_3|\phi_3\rangle \right) = 0$$

Remember that $\langle\phi_i|\phi_i\rangle = 1$ and that in the Hückel approximation, $\langle\phi_i|\phi_j\rangle = 0$ for $i \neq j$.

$$\frac{1}{\sqrt{2}}(c_2 - c_3) = 0$$

$$c_2 = c_3$$

Using this result and forcing orthogonality between ψ_1 and ψ_3,

$$\left\langle \frac{1}{\sqrt{3}}(\phi_1 + \phi_2 + \phi_3) \middle| c_1\phi_1 + c_2\phi_2 + c_2\phi_3 \right\rangle = 0$$

$$\frac{1}{\sqrt{3}}(c_1 + c_2 + c_2) = 0$$

$$c_1 = -2c_2$$

Normalizing ψ_3,

$$(-2c_2)^2 + c_2{}^2 + c_2{}^2 = 1$$

$$c_2 = \frac{1}{\sqrt{6}}, \text{ so } c_3 = \frac{1}{\sqrt{6}} \text{ and } c_1 = -\frac{2}{\sqrt{6}}$$

Summarizing,

$$\psi_1 = \frac{1}{\sqrt{3}}(\phi_1 + \phi_2 + \phi_3), \; E = H_{AA} + 2H_{AB} \; (E = \alpha + 2\beta)$$

$$\psi_2 = \frac{1}{\sqrt{2}}(\phi_2 - \phi_3), \; E = H_{AA} - H_{AB} \; (E = \alpha - \beta)$$

$$\psi_3 = \frac{1}{\sqrt{6}}(-2\phi_1 + \phi_2 + \phi_3), \; E = H_{AA} - H_{AB} \; (E = \alpha - \beta)$$

15. Before looking at the experimental data, let's consider what we would expect it to look like for compounds **1**-**3**. The PES data will give us energies of filled orbitals, in this case the π MOs. Compound **3** has one π bond and should have one π MO. The plane of the alkene C's and the attached atoms is a molecular symmetry plane, and the π MO is A with respect to this plane. Diene **1** has two π MOs, one S and one A with respect to the analogous plane. Mixing of these MOs to form the three π MOs of **3** is straightforward: only the A orbitals can mix. The result should be a mixing diagram like this:

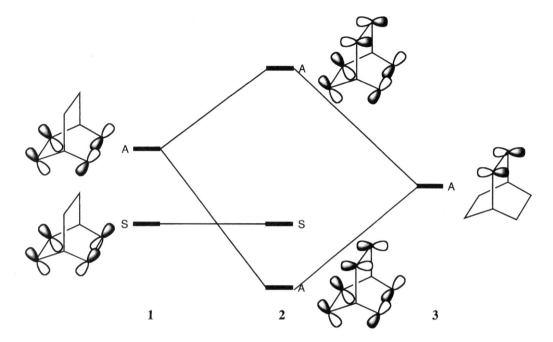

If we plot the diagram to scale with the experimental energies, we find that the diagram looks much like the one above.

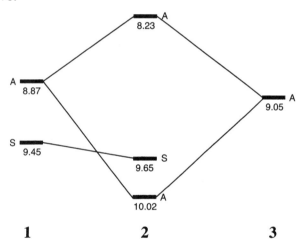

However, a similar plot for **4-6** suggests that the S and A MOs appear in the opposite order:

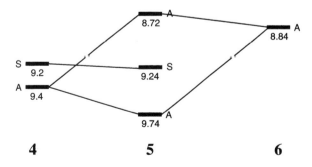

The reversal is has been attributed to a more efficient through-bond coupling involving the C–C bonds to the third bridge.

16. If the molecule is planar, then both C–H bonds significantly overlap with the adjacent p orbitals. If the CH$_2$ group comes out of the plane as shown, one of the C–H bonds aligns better with the p orbitals at the expense of the other. So the net through-bond coupling effect is similar.

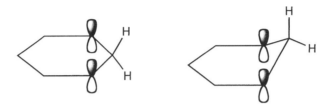

17. The determinants below use the following labeling scheme:

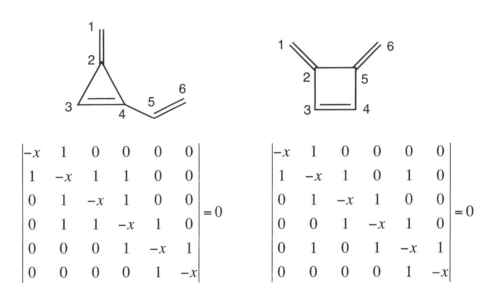

$$
\begin{vmatrix}
-x & 1 & 0 & 0 & 0 & 0 \\
1 & -x & 1 & 1 & 0 & 0 \\
0 & 1 & -x & 1 & 0 & 0 \\
0 & 1 & 1 & -x & 1 & 0 \\
0 & 0 & 0 & 1 & -x & 1 \\
0 & 0 & 0 & 0 & 1 & -x
\end{vmatrix} = 0
\qquad
\begin{vmatrix}
-x & 1 & 0 & 0 & 0 & 0 \\
1 & -x & 1 & 0 & 1 & 0 \\
0 & 1 & -x & 1 & 0 & 0 \\
0 & 0 & 1 & -x & 1 & 0 \\
0 & 1 & 0 & 1 & -x & 1 \\
0 & 0 & 0 & 0 & 1 & -x
\end{vmatrix} = 0
$$

18. The π system of each of the molecules can be viewed as coming from a joining of benzyl and allyl. The NBMOs, being the MOs associated with the unpaired electrons, give us the most insight into these coupling processes. More stable π systems will arise when there is higher overlap due to the junction. (It is important to realize that though we are analyzing the π systems as coming from benzyl + allyl, the new σ bonds would require the loss of H atoms. Actual coupling of benzyl + allyl would give a molecule with only one new σ bond and two sp^3 C atoms.)

First, a reminder of what the benzyl and allyl NBMOs look like:

The NBMO overlaps can be easily computed from the interacting AO coefficients. The braket notation is a convenient way to set up the calculation. The compounds are shown below in order of highest to lowest junction overlap.

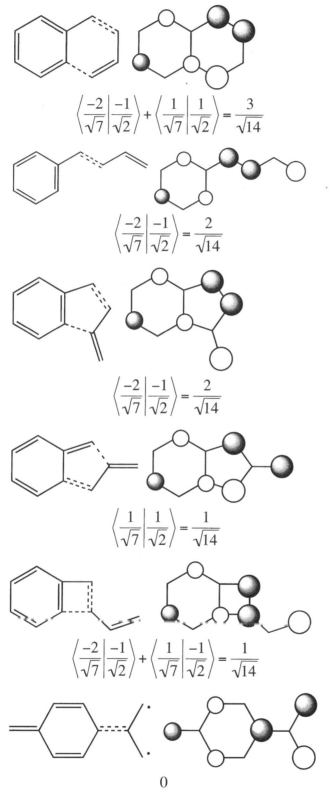

$$\left\langle \frac{-2}{\sqrt{7}} \middle| \frac{-1}{\sqrt{2}} \right\rangle + \left\langle \frac{1}{\sqrt{7}} \middle| \frac{1}{\sqrt{2}} \right\rangle = \frac{3}{\sqrt{14}}$$

Both junctions have good positive overlap. (A new aromatic ring is formed, and the new ring orbital resembles a bonding orbital of benzene).

$$\left\langle \frac{-2}{\sqrt{7}} \middle| \frac{-1}{\sqrt{2}} \right\rangle = \frac{2}{\sqrt{14}}$$

A strong, new π bond is formed between the C's with the largest coefficients.

$$\left\langle \frac{-2}{\sqrt{7}} \middle| \frac{-1}{\sqrt{2}} \right\rangle = \frac{2}{\sqrt{14}}$$

Though similar to the last one, nothing is gained by the second connection to the center of the allyl unit.

$$\left\langle \frac{1}{\sqrt{7}} \middle| \frac{1}{\sqrt{2}} \right\rangle = \frac{1}{\sqrt{14}}$$

Though similar to the previous isomer, this one is not as favorable. The larger benzyl coefficient is paired with the zero coefficient of allyl.

$$\left\langle \frac{-2}{\sqrt{7}} \middle| \frac{-1}{\sqrt{2}} \right\rangle + \left\langle \frac{1}{\sqrt{7}} \middle| \frac{-1}{\sqrt{2}} \right\rangle = \frac{1}{\sqrt{14}}$$

Overlap at the new junctions give cancellation of overlap, though not complete cancellation due to the different benzyl coefficients. The new ring is antiaromatic.

$$0$$

This biradical gains nothing over benzyl + allyl, since the only connection involves the central C of allyl, having a zero coefficient.

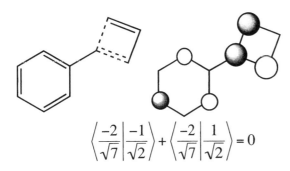

In the formation of this antiaromatic ring, there is complete cancellation of overlap, suggesting that this junction is comparable to the formation of the biradical above.

$$\left\langle \frac{-2}{\sqrt{7}} \middle| \frac{-1}{\sqrt{2}} \right\rangle + \left\langle \frac{-2}{\sqrt{7}} \middle| \frac{1}{\sqrt{2}} \right\rangle = 0$$

19. The calculated energies below match those in Figure 14.12. Note that $N = 5$ and that i goes from 0 to ±2. Since the cosine function does not depend on sign, only three different energies are obtained.

$$\varepsilon_0 = \alpha + 2\beta \cos\left(\frac{0}{5}\right) = \alpha + 2\beta$$

$$\varepsilon_1 = \alpha + 2\beta \cos\left(\frac{2\pi}{5}\right) = \alpha + 2\beta(0.3090) = \alpha + 0.618\beta$$

$$\varepsilon_2 = \alpha + 2\beta \cos\left(\frac{4\pi}{5}\right) = \alpha + 2\beta(-0.8090) = \alpha - 1.618\beta$$

20. Starting with the observation that the higher energy degenerate orbitals should have two nodal planes, we can arrive at the general shapes of the MOs. Our task is then to use the rules to obtain the three coefficients.

 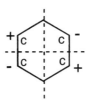

From rule 1 (normalization), we can obtain c directly:

$$4c^2 = 1, \text{ so } c = \frac{1}{2}$$

Normalization of the other MO gives

$$2a^2 + 4b^2 = 1$$

Rule 3 gives

$$a^2 = b^2 + c^2 = b^2 + \frac{1}{4}$$

Plugging this into the normalization equation gives

$$2\left(b^2 + \frac{1}{4}\right) + 4b^2 = 1$$

$$6b^2 = \frac{1}{2}$$

$$b = \frac{1}{\sqrt{12}}$$

Using the normalization equation to solve for a:

$$2a^2 + 4\left(\frac{1}{\sqrt{12}}\right)^2 = 1$$

$$2a^2 = \frac{2}{3}$$

$$a = \frac{1}{\sqrt{3}} = \frac{2}{\sqrt{12}}$$

So the two MOs are

$$\psi = \frac{1}{\sqrt{12}}\left(2\phi_1 - \phi_2 - \phi_3 + 2\phi_4 - \phi_5 - \phi_6\right) \text{ and } \psi = \frac{1}{2}\left(\phi_2 - \phi_3 + \phi_5 - \phi_6\right)$$

21. First order mixing is all that is needed in this case, since symmetry must be preserved. Note that two NBMOs are formed, since the allyl NBMOs are joined at carbon atoms that have coefficients equal to zero.

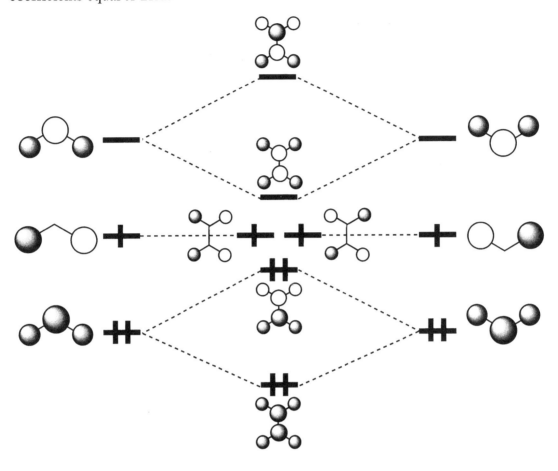

22. Since the unpaired electrons are not interacting in the transition state, we can calculate the π energy of the transition state as the sum of the energies for methyl and allyl radicals. Subtracting the π energy of butadiene should give us the activation energy, since there are no changes in the σ bonding. The π energies are obtained by summing the energies for all π electrons.

$$E = 2(\alpha + 1.618\beta) + 2(\alpha + 0.618\beta) \qquad E = \alpha + 2(\alpha + \sqrt{2}\beta) + \alpha$$

$$E = 4\alpha + 4.472\beta \qquad\qquad E = 4\alpha + 2\sqrt{2}\beta = 4\alpha + 2.828\beta$$

Activation energy $= (4\alpha + 2.828\beta) - (4\alpha + 4.472\beta) = -1.644\beta$

Note that β is negative, so the activation energy is a positive number.

23. Because electrons are fermions, the total wavefunction must be antisymmetric with respect to the exchange of any two electrons. This gives rise naturally to determinantal wavefunctions. If we put three electrons into an orbital, two will necessarily have the same spin. This pair has identical space (in same orbital) and spin characteristics, and so in a Slater determinant comprised of spin orbitals, they would produce two columns that are identical. A determinant with two identical columns has a value of zero, and so placement of three electrons into an orbital is not possible.

24. The semi-empirical methods consider valence orbitals only in the basis set, while ab initio methods consider all orbitals up to and including the valence orbitals even for the most minimal basis set. For C, this simply means the 1s orbital is neglected, a small savings. However, for Si by comparison, the semi-empirical approach neglects 1s, 2s, and 2p orbitals, all of which are included in ab initio approaches. This is a significant difference.

25. Eq. 14.53, with $k = 1$, would reduce to

$$H_{ij} = \frac{H_{ii} + H_{jj}}{2} S_{ij}$$

For simplicity, let's consider a symmetric interaction, for which $H_{ii} = H_{jj}$. In this case,

$$H_{ij} = H_{ii}S$$

Using Eq. 14.46, we can calculate the energies of the MOs that result from the interaction:

$$\varepsilon_1 = \frac{H_{ii} + H_{ij}}{1 + S} = \frac{H_{ii} + H_{ii}S}{1 + S} = \frac{H_{ii}(1 + S)}{1 + S} = H_{ii}$$

$$\varepsilon_2 = \frac{H_{ii} - H_{ij}}{1 - S} = \frac{H_{ii} - H_{ii}S}{1 - S} = \frac{H_{ii}(1 - S)}{1 - S} = H_{ii}$$

So the interaction has no energetic consequence: the MOs resulting from the interaction have the same energy as the starting orbitals. It is thus clear that a bonding interaction will require k to be greater than 1, such that H_{ij} dominates $H_{ii}S$. In other words, the overlap of orbitals must result in a lowering of an electron's energy, corresponding to an increase in the magnitude of H_{ij}. (Remember that H_{ii} and H_{ij} are negative, representing stabilization relative to a free electron.)

26. Two reasons for a shorter C–C bond in ethyl cation:
   (1) The bond in ethyl cation is an sp^2–sp^3 bond, while the bond in ethane is sp^3–sp^3.
   (2) Hyperconjugation gives the bond some π-bonding character:

27. All of the bond lengths can be rationalized in reference to the interaction depicted in Figure 14.24. The very short cyclopropyl–CH$_2$ bond is a result of the π-type bonding interaction. The same interaction leads to a partial depopulation of the Walsh orbital below as it is mixed into the unfilled, antibonding orbital. This depopulation weakens bonds that are bonding in this Walsh orbital and strengthens bonds that are antibonding in this orbital. Therefore, the bonds adjacent to the substituent are lengthened, while the bond opposite the substituent is shortened.

28. We can first determine the rest of the absolute values of the coefficients. Here is what we know and what needs to be determined:

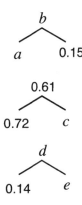

By using Rule 2 (that each atomic orbital must be "used up"), we can calculate a:

$$a^2 + (0.72)^2 + (0.14)^2 = 1$$

$$a = 0.68$$

Now we can obtain both b and c by using Rule 1 (normalization):

$$(0.68)^2 + b^2 + (0.15)^2 = 1 \qquad\qquad (0.72)^2 + (0.61)^2 + c^2 = 1$$

$$b = 0.72 \qquad\qquad\qquad\qquad c = 0.33$$

Finally, d and e are determined by using Rule 2:

$$(0.72)^2 + (0.61)^2 + d^2 = 1 \qquad\qquad (0.15)^2 + (0.33)^2 + e^2 = 1$$

$$d = 0.33 \qquad\qquad\qquad\qquad e = 0.93$$

As a double check, we can verify that the lowest energy MO is normalized:

$$(0.14)^2 + (0.33)^2 + (0.93)^2 = 1$$

$0.993 = 1$, so the MO is normalized (within rounding error).

The coefficient absolute values are:

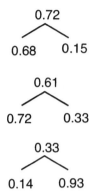

A qualitative MO diagram for vinyl alcohol is obtained by mixing the π orbitals of ethylene with the p_π orbital of O. The electronegativity of O results in a lowering of the p_π orbital, such that it mixes more strongly with the π orbital of ethylene rather than the π*. The calculated coefficients show that the O p orbital actually lies at lower energy than the ethylene π orbital (at least at the HMO level).

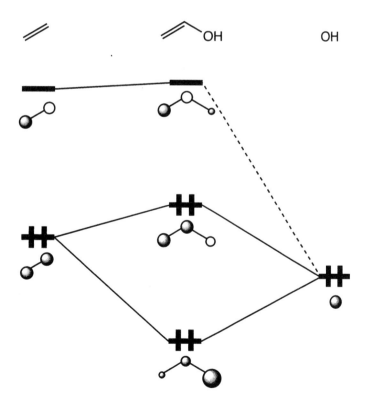

This diagram is consistent with the calculated coefficients, so the relative signs may be taken directly. As expected, the lowest MO has no nodes, the middle MO has one node, and the highest MO has two nodes.

$$-0.72$$

$$0.68 \quad OH \quad 0.15$$

$$0.61$$

$$0.72 \quad OH \quad -0.33$$

$$0.33$$

$$0.14 \quad OH \quad 0.93$$

29. The mixing responsible for this σ bond lengthening must involve the butadiene π system. In order for the mixing to cause any observable bond lengthening, either the σ orbital must be depopulated or the σ* orbital must be populated (or both). This requires a mixing of filled and empty orbitals, *i.e.*, either σ-π* or σ*-π. Considering the σ-type orbitals to be the cyclopropane Walsh orbitals, there are two choices for σ and one for σ*. Since cyclopropane is a good donor and poor acceptor, we can focus on the σ orbitals. Only one σ orbital can interact with the butadiene LUMO, both of which are symmetric with respect to the molecular symmetry plane. So the important interaction is σ-π*, resulting in a partial depopulation of the σ orbital:

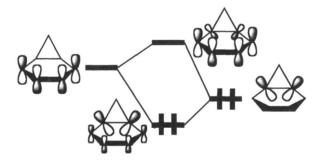

30. a. The lone pair on N can participate in a strong π interaction with the C p orbital:

This interaction serves to increase the energy gap between the two "non-bonding" carbene orbitals, raising the energy of the triplet state, which would have one electron in the higher orbital.

b. Increasing the electronegativity of the substituent lowers the energy of the lone pair orbital discussed in Part A. This increases the energy gap and so makes the orbital mixing weaker. Thus, the strong singlet preference of aminomethylene progressively weakens on replacing NH_2 with OH and F.

31. a. In the trigonal bipyramidal geometry, the C 2p orbital interacts strongly with the antisymmetric combination of the H 1s orbitals:

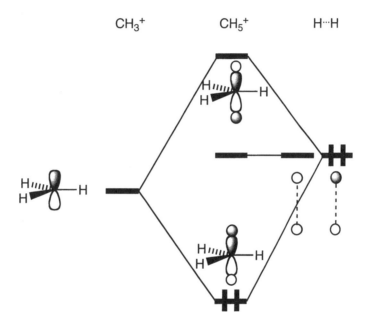

b. In the alternative geometry, the C $2sp^3$ orbital interacts with the H_2 σ orbital:

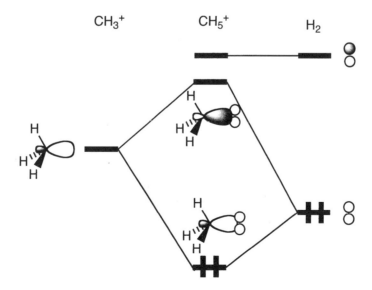

c. The interaction shown in the trigonal bipyramidal case is presumably stronger, due to a smaller energy gap, but the H_2 also starts with a broken bond in this case. Therefore, a stronger interaction is required for the electrons to achieve the same degree of stabilization. In the bottom case, the H's can still bond to each other while they are both bonding to C. One effect that goes beyond the diagrams shown is the notion of H exchange. In the trigonal bipyramidal geometry, the adjacent H–C–H angles vary from 90 to 120°, while the spread is much less in the latter geometry. This might allow the charge to more effectively delocalize to all five H's without requiring a change in geometry. (The highlight on CH_5^+ in Section 1.4.1 seems to support this view.)

32. While there are indeed four π MOs for the tetraene **1**, two of them are degenerate. Even though these two come from different mixings in the diagram below, one arising as an antibonding combination of the lower butadiene MOs and the other as a bonding combination of the upper butadiene MOs, the fourfold symmetry of **1** requires that these MOs be degenerate. Each of these orbitals possesses only two-fold symmetry on its own. (Note that the relative coefficients shown for the degenerate MOs, reflecting their butadiene origins, can be adjusted to make the two orbitals look the same. Note also that butadiene is a good, though not perfect model for **1**. The bond angles are quite different, but the angle effect on the π MO energies should be fairly small.)

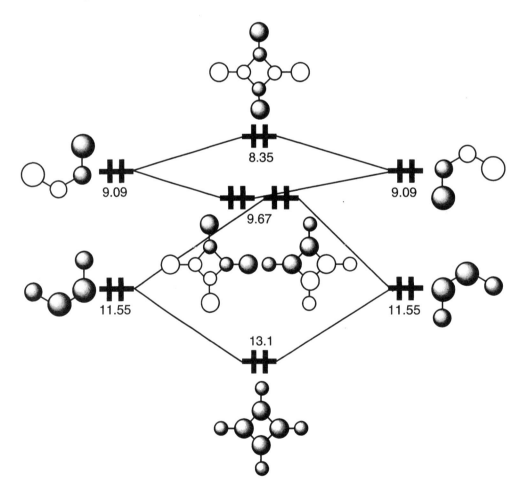

33. Deformation of planar methyl to a T-shaped geometry primarily affects the π(CH₃) MOs.

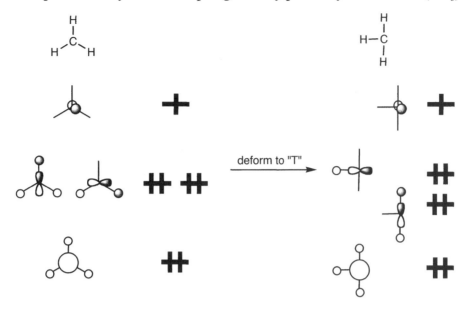

We can now construct planar ethane:

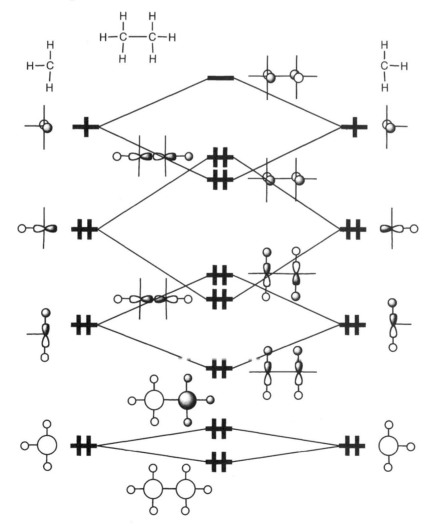

The most interesting feature of this diagram is that the only stabilizing interaction involving the C-C bond is a π interaction. In other words, the C–C bond in planar ethane has a π bond but not a σ bond, at least in an energetic sense. Unlike planar methane, planar ethane has enough electrons in C–H bonding orbitals to have two electrons per bond. This is made possible by the use of the π symmetry orbitals for the C–C π bond. We should not, however, conclude that this is a stable geometry. One problem, in addition to the non-optimum C hybridization, is that some of the C–H bonding MOs have significant C–C and H··H antibonding character.

34. The fact that one ionization of bicyclopentane has the same energy (11.2 eV) as one for bicyclopentene suggests that this Walsh orbital has the wrong symmetry to mix with the alkene π MO. Since the π MO is symmetric with respect to the molecular mirror plane of bicyclopentene, the lower Walsh orbital must be the antisymmetric one. We obtain the following mixing diagram:

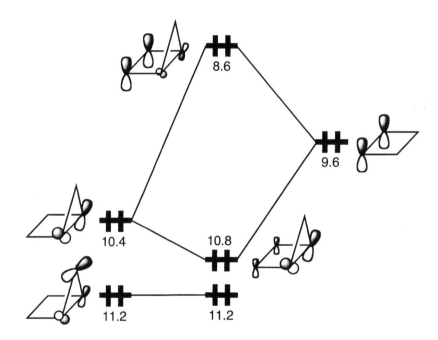

35. The MO mixing diagram below for ethylene is an abridged version of Figure 1.14, showing the interactions that are most important for the present comparison.

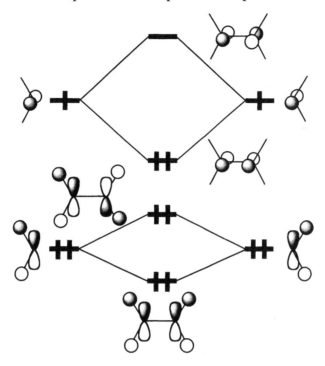

As shown, there are two π interactions that can lead to conformational preferences with respect to rotation about the C–C bond. The filled-filled $\pi(CH_2)$–$\pi(CH_2)$ interaction is slightly repulsive, but the p_π–p_π interaction is strongly bonding, leading to a planar geometry for ethylene.

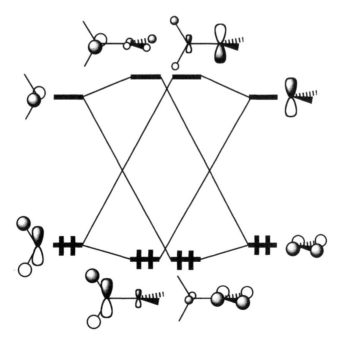

The diagram for planar H_2B–BH_2 would be the same, except that there are two fewer electrons. Without the p_π–p_π bonding interaction, the only π interaction, $\pi(CH_2)$–$\pi(CH_2)$, would be repulsive. A 90° rotation not only removes this filled-filled interaction but also allows for two filled-empty, bonding π interactions. These hyperconjugative interactions are weaker than the π bonding in ethylene, due to the large energy gap. They are, however, strong enough to provide a significant rotational barrier, as found by numerous theoretical studies. (B_2H_4 is an unstable molecule and has not been isolated.)

36. a. One of the *p*-xylene MOs has nearly the same energy as the degenerate HOMO of benzene, so we can presume that to be the one that has coefficients of zero at the substituted C's. The other MO is raised in energy, due to the electron-donating nature of the Me groups. (In MO terms, this is due to interactions with $\pi(CH_3)$ orbitals.)

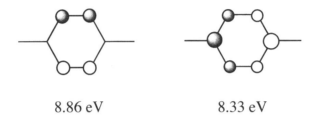

8.86 eV   8.33 eV

b. When the Me groups and the attached C's are pulled out of the plane, two changes take place that affect only the second orbital: (1) twisting of the dihedral angles at the substituted C, reducing p-p overlap, and (2) partial pyramidalization of the substituted C's, also reducing overlap. (The latter effect is exaggerated in the diagram below.) Thus, we should expect the first MO to stay nearly the same in energy, while the second MO should increase in energy:

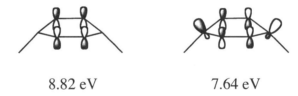

8.82 eV   7.64 eV

c. Joining two of these π systems in a face-to-face manner will allow each of these MOs to form bonding and antibonding combinations. By plotting the fragment and final MOs on the same energy scale, we can readily deduce the mixing pattern and assignments. Each MO of compound **1** is also shown from an alternative perspective: from the left side of the original structure.

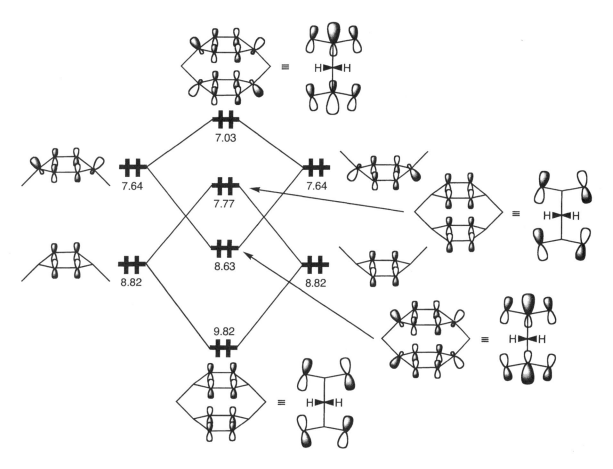

37. HMO is an NDO (neglect of differential overlap) method; in other words, the overlap of different, even adjacent, orbitals is assumed to be zero. Therefore, normalization of an MO can be done by considering only the component AOs: the sum of the squares of the AO coefficients must equal 1. In EHT, overlap is not neglected, so the normalization calculation is not as simple and different coefficients for the AOs may be obtained. For the fully π bonding MO of benzene, the adjacent p orbitals have positive, reinforcing overlap, meaning that the AO coefficients must be smaller than in HMO in order to get correct normalization: $\langle \psi | \psi \rangle = 1$. The coefficient obtained, 0.33, is indeed less than that obtained by HMO, 0.41. In the fully antibonding MO, adjacent p orbitals have negative, canceling overlap, requiring a larger coefficient (0.54) than that from HMO (0.41).

38. All five bond length differences can be explained by considering the interaction of one of the HOMO Walsh orbitals of the cyclopropane and the LUMO of cyclopentadiene. Note that the HOMO of cyclopentadiene has the wrong symmetry to interact with any of the Walsh orbitals.

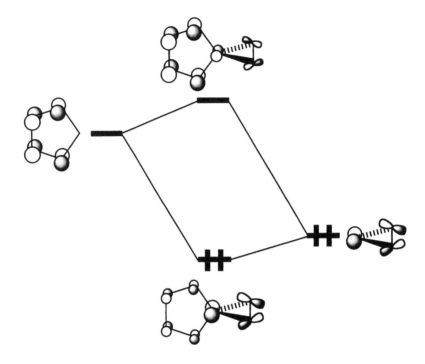

We can rationalize the changes in the cyclopentadiene bond lengths as arising from a partial occupation of the empty orbital as it mixes into the filled MO. The overlap at bond a is positive, so it should be shortened; the observed effect is very small. The overlap at bond b is negative, so it should be lengthened, as is observed. The changes in the cyclopropane come from depopulation of the filled Walsh orbital. The overlap at bond d is positive, so it should be lengthened, as is observed. The overlap at bond e is negative, so it should be shortened, as is observed. Bond c has positive overlap in the filled MO, resulting in a shortening relative to cyclopentadiene.

39. The obvious difference between these isomers is that direct (through space) overlap between the π orbitals is possible only in the second one (the *syn* isomer). As with any mixing, this should cause one combination to drop in energy and the other to rise. So it might seem strange that this isomer actually has a smaller splitting between the π MOs. The explanation both for this and for the large splitting observed for the *anti* isomer is through-bond coupling, involving C–C bonds in the central ring. This effect counters the through-space effect, bringing the MOs back together for the *syn* isomer and spreading them apart for the *anti* isomer. Note that it is unclear from the data whether the through space or through bond interactions are stronger, *i.e.*, whether the S or A MO is the lower one in the *syn* isomer.

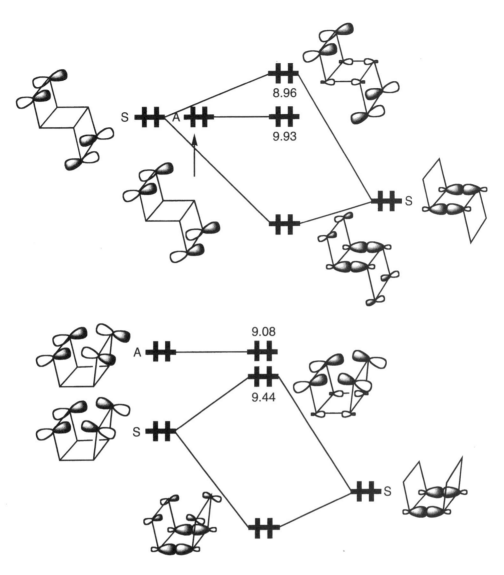

40. The through-bond coupling involves $\pi(CH_2)$ orbitals, two in cyclobutanediyl and one in cyclopentanediyl:

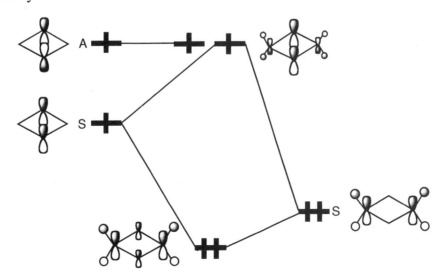

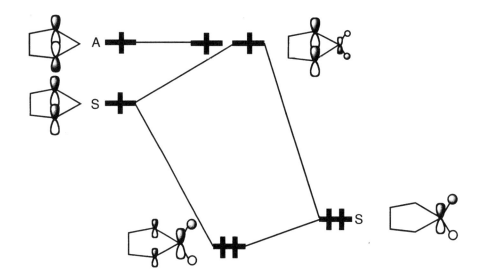

The through-space interaction is smaller in cyclopentanediyl because of the larger CCC angle of the five-membered ring, which puts the orbitals further apart. However, the through-bond mixing is also smaller in cyclopentanediyl because there is only one $\pi(CH_2)$ orbital to mix with, vs. two in cyclobutanediyl. The two effects roughly cancel each other, so in both cases the two NBMOs are very nearly degenerate.

Ignoring the participation of $\pi(CH_2)$, the S and A orbitals of both cyclobutanediyl and cyclopentanediyl are disjoint. Starting with the orbitals shown on the left in the above diagrams and mixing them equally in linear combinations, we obtain MOs that clearly have no non-zero coefficient on common atoms.

The participation of $\pi(CH_2)$ changes this. The appropriate combinations that cancel out the p AOs give MOs that have contributions from $\pi(CH_2)$:

Therefore, the NBMOs are nondisjoint.

41. The interaction between cyclopropane and methyl cation in the Coates geometry has a mirror plane that is horizontal in the orientation shown. The methyl p orbital is symmetric with respect to the plane, so the only cyclopropane Walsh orbital that can interact with it is the symmetric one, which happens to be filled. The interaction is therefore favorable.

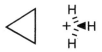

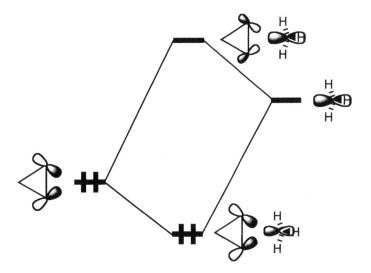

42. The diagram below shows that the two empty NBMOs of CH^+ can mix with the half-filled NBMOs of cyclobutadiene, leading to highly stabilizing interactions in which the non-bonding electrons are placed in bonding MOs. These bonding MOs can also accept two electrons that would otherwise reside in an antibonding MO of SS symmetry. Therefore, $(CH)_5^+$ is stabilized relative to the fragments. The symmetry labels refer to the mirror planes shown at the upper left. (For simplicity, the C–H bonding orbital of CH^+ is shown as non-interacting. In reality, it will mix with the other SS orbitals, but this mixing does not affect the question of whether the association of these fragments is stabilizing.)

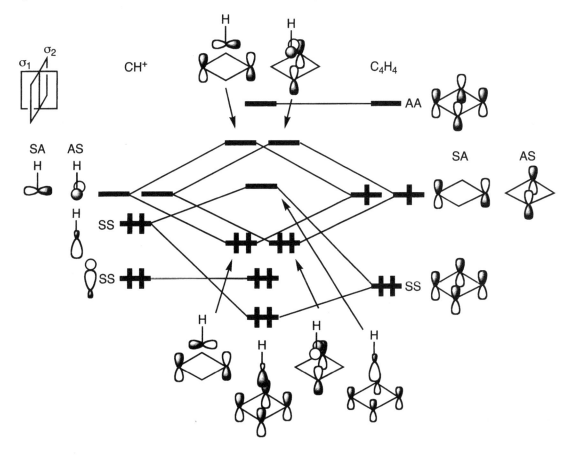

43. Even without considering specific orbitals, we can reason that cyclobutadiene should be better than benzene in general at forming complexes with other fragments. Cyclobutadiene has two non-bonding electrons that can be stabilized in the interaction, leading to a stabilization of the complex relative to the fragments. All electrons in benzene are already in bonding MOs, so stabilization of a complex, though possible, would likely be less.

The diagram below is slightly simplified relative to the one in exercise 42, showing only the key interactions. There is a stabilizing interaction between the HOMOs of benzene and the empty p orbitals of CH^+. However, this should be weaker than the interaction between CH^+ and cyclobutadiene because of the poorer overlap expected with the large six-membered ring and the larger energy gap. Also, there is a destabilizing closed-shell interaction involving the SS orbitals. The diagram thus suggests that the complex might be slightly stabilized relative to the fragments, but there is certainly not a strong stabilizing interaction.

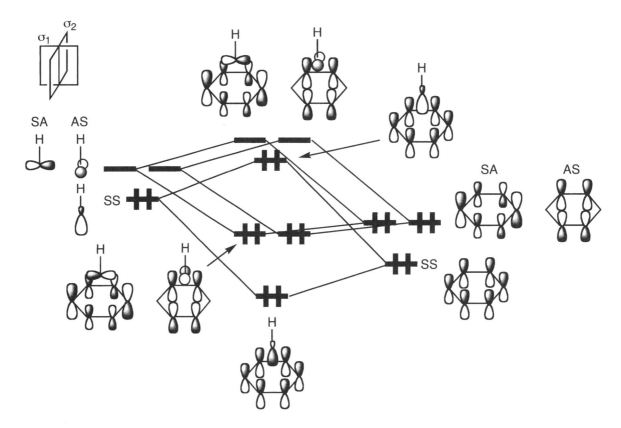

44. These MOs are the antibonding combinations of the M-L σ bonds, with primary character on the metal center. Population of these orbitals will weaken the bonds between the metal and ligands, destabilizing the complex.

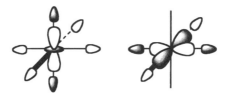

45. The metal AOs and the benzene π orbitals are shown below with their symmetries with respect to three mirror planes.

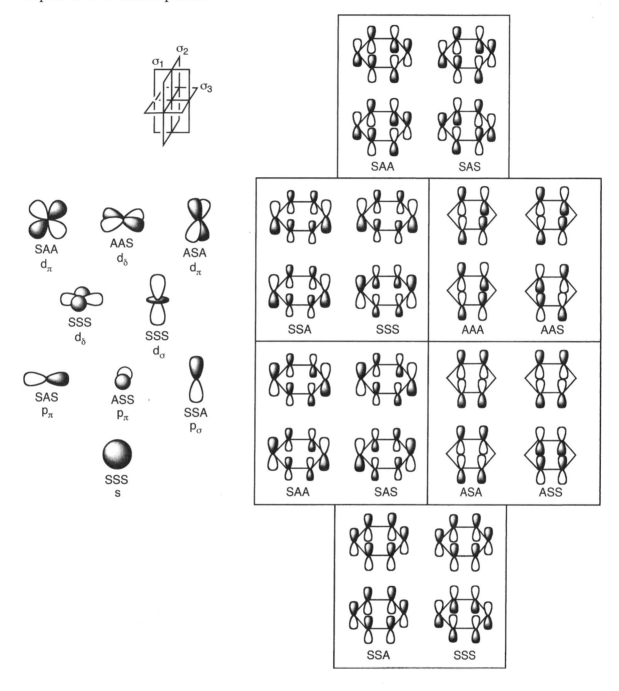

Rather than draw a complete mixing diagram, the benzene orbitals are represented below along with their symmetries and metal AO mixing partners. The strongest interactions are the two involving d_π interactions, since these have both high overlaps and low energy gaps. The bonding combinations for these are shown. Note that the molecule actually has more symmetry than just the three mirror planes (including a C_6 axis and four more vertical mirror planes). This is why d_σ and d_δ metal AOs can have identical symmetry labels. The highest benzene MOs do not mix significantly with the metal AOs, due to cancellation of overlap arising from the three nodal planes.

	SAA	SAS
	p_σ	d_δ		d_δ
	SSA	SSS	AAA	AAS
d_π	p_π	d_π	p_π
SAA	SAS	ASA	ASS
	p_σ	s, d_σ
	SSA	SSS

SAA   ASA

The least important d orbitals in the bonding are the two of δ symmetry. Since these orbitals lie parallel to the benzene rings, the overlap is low.

46. Metal fragments that will form triple bonds to methine should be isolobal with methine, since methine forms triple bonds with itself to form acetylene. Therefore, any fragment represented in the third row of the table should form a triple bond with CH. Some examples of the resulting complexes are shown.

$(d^5\text{-ML}_5)$ $(d^7\text{-ML}_4)$ $(d^{10}\text{-CpM})$ $(d^9\text{-ML}_3)$

47. If the fragment is getting one fewer electron from its ligands, it will need an additional electron of its own in order to remain in the same isolobal group. So the entries in the table below each have one more d electron than in Table 14.1.

Isolobal relationships when one ligand (X) contributes one electron and the others (L) contribute two. (Note that the d electron count shown is that of the M atom before coordination.)

Organic groups	Metal fragments				
methyl	$d^{10}\text{-MXL}_3$	$d^8\text{MXL}_4$	$d^6\text{MXL}_5$	$d^4\text{MXL}_6$	$d^9\text{CpMXL}$
methylene	$d^9\text{MXL}_3$	$d^7\text{MXL}_4$	$d^5\text{MXL}_5$	$d^3\text{MXL}_6$	$d^{10}\text{CpMX}$
methine	$d^{10}\text{MXL}_2$	$d^8\text{MXL}_3$	$d^6\text{MXL}_4$	$d^4\text{MXL}_5$	$d^2\text{MXL}_6$

48. By definition, the transition state is the highest energy point on the lowest energy path between reactants and products. If the "transition state" is at a maximum with respect to more than one dimension, there will always be an alternative, lower energy path in which one of the maxima is replaced by a minimum. This is very familiar to us on a two-dimensional surface: the lowest path from one valley to the next is always through a mountain pass and never over a summit!

49. To see why only one- and two-electron integrals need to be considered, why the one-electron integrals always involve the same orbital, and why the two-electron integrals always involve only the same two orbitals, let's look at a three-electron system. Remember that the n HF equations arise from an $n \times n$ determinantal wavefunction used with the Hamiltonian for the entire system. The antisymmetrized Ψ for a three-electron system is given in Eq. A, where we simplify the notation such that $a = \psi_a$. We also simplify the Hamiltonian for a three-electron system to that given in Eq. B, were h_i stands for the core portion (terms ③ and ④ from Figure 14.3) and $1/r_{ij}$ stands for the electron-electron interactions (term ⑤). We compute the energy in Eq. C, in which the overlap integral in the denominator equals 1, since Ψ is a normalized wavefunction. A further simplification is used in this equation: only one of the six terms of Ψ is included in the bra portion of the integral and the result is multiplied by 6. (After finishing the solution, you might convince yourself that each of the six terms would give an equivalent result.)

Eq. A

$$\Psi = \psi_a(1)\psi_b(2)\psi_c(3) - \psi_a(1)\psi_c(2)\psi_b(3) - \psi_b(1)\psi_a(2)\psi_c(3) + \psi_b(1)\psi_c(2)\psi_a(3)$$
$$+ \psi_c(1)\psi_a(2)\psi_b(3) - \psi_c(1)\psi_b(2)\psi_a(3)$$
$$= a(1)b(2)c(3) - a(1)c(2)b(3) - b(1)a(2)c(3) + b(1)c(2)a(3) + c(1)a(2)b(3) - c(1)b(2)a(3)$$

Eq. B

$$H(1,2,3) = h_1 + h_2 + h_3 + \frac{1}{r_{12}} + \frac{1}{r_{13}} + \frac{1}{r_{23}}$$

Eq. C

$$E = \frac{\langle \Psi | H | \Psi \rangle}{\langle \Psi | \Psi \rangle} = \langle \Psi | H | \Psi \rangle$$

$$= 6 \langle a(1)b(2)c(3) | h_1 + h_2 + h_3 + \frac{1}{r_{12}} + \frac{1}{r_{13}} + \frac{1}{r_{23}} | a(1)b(2)c(3) - a(1)c(2)b(3) - b(1)a(2)c(3)$$
$$+ b(1)c(2)a(3) + c(1)a(2)b(3) - c(1)b(2)a(3) \rangle$$

Let's first just look at all the terms involving h_1, and factor out those orbitals that do not have a dependence upon electron 1, grouping them within integrals over each electron (Eq. D).

Since the orbitals a, b, and c are orthogonal and normalized, any term such as $\langle b(2)|c(2)\rangle$ will equal zero and any term such as $\langle b(2)|b(2)\rangle$ will equal 1. Only the first integral of Eq. D survives. Similar results are obtained for the h_2 and h_3 containing terms, and hence all the core integrals for a multiple electron system consist solely of "one-electron" integrals with the same wavefunction in the bra and ket portion of the integral.

Eq. D

$$
\begin{aligned}
E(h_1 \text{ terms}) = 6\big(&\langle a(1)|h_1|a(1)\rangle\langle b(2)|b(2)\rangle\langle c(3)|c(3)\rangle - \langle a(1)|h_1|a(1)\rangle\langle b(2)|c(2)\rangle\langle c(3)|b(3)\rangle \\
&- \langle a(1)|h_1|b(1)\rangle\langle b(2)|a(2)\rangle\langle c(3)|c(3)\rangle + \langle a(1)|h_1|b(1)\rangle\langle b(2)|c(2)\rangle\langle c(3)|a(3)\rangle \\
&+ \langle a(1)|h_1|c(1)\rangle\langle b(2)|a(2)\rangle\langle c(3)|b(3)\rangle - \langle a(1)|h_1|c(1)\rangle\langle b(2)|b(2)\rangle\langle c(3)|a(3)\rangle\big) \\
= 6&\langle a(1)|h_1|a(1)\rangle
\end{aligned}
$$

Now let's consider the $1/r_{12}$ terms. We once again factor out the wavefunctions that do not contain electrons 1 and 2 from the $1/r_{12}$ integrals, obtaining Eq. E. Once again, since the wavefunctions are orthogonal and normalized, only the first and third terms survive. Note that the surviving terms are the Coulomb and Exchange integrals as defined above (Eqs. 14.29 and 14.30), and that they only consist of two-electron interactions with the same two orbitals. Hence, no integrals of the type $\langle ij|1/r|kl\rangle$ contribute to the HF analysis. The same occurs for the $1/r_{13}$ and $1/r_{23}$ terms.

Eq. E

$$
\begin{aligned}
E(1/r_{12} \text{ terms}) = 6\Big(&\langle a(1)b(2)|\tfrac{1}{r_{12}}|a(1)b(2)\rangle\langle c(3)|c(3)\rangle - \langle a(1)b(2)|\tfrac{1}{r_{12}}|a(1)c(2)\rangle\langle c(3)|b(3)\rangle \\
&- \langle a(1)b(2)|\tfrac{1}{r_{12}}|b(1)a(2)\rangle\langle c(3)|c(3)\rangle + \langle a(1)b(2)|\tfrac{1}{r_{12}}|b(1)c(2)\rangle\langle c(3)|a(3)\rangle \\
&+ \langle a(1)b(2)|\tfrac{1}{r_{12}}|c(1)a(2)\rangle\langle c(3)|b(3)\rangle - \langle a(1)b(2)|\tfrac{1}{r_{12}}|c(1)b(2)\rangle\langle c(3)|a(3)\rangle\Big) \\
= 6\Big(&\langle a(1)b(2)|\tfrac{1}{r_{12}}|a(1)b(2)\rangle - \langle a(1)b(2)|\tfrac{1}{r_{12}}|b(1)a(2)\rangle\Big)
\end{aligned}
$$

These results can be generalized for multi-electron systems. In all cases, we find that the integrals can be reduced to one and two electron integrals of the form H_{ii} and J_{ij} and K_{ij}. This is because the orbitals involving other electrons are factored out of the Hamiltonian-containing integrals, and in composite give either a factor of 1 or 0, causing the overall product of integrals to survive or vanish. Another way to understand this is that the Hamiltonian is a one or two electron operator. It is solely concerned with the kinetic energy and attraction of each single electron with all the nuclei, and the interactions between any two electrons. Hence, it makes good sense that only one and two electrons integrals should result.

15

Thermal Pericyclic Reactions

S O L U T I O N S   T O   E X E R C I S E S

1. The $_\pi 2_s + _\pi 2_a$ cycloaddition of ethylene differs from the $_\pi 2_s + _\pi 2_s$ cycloaddition not only in the symmetry elements that are preserved through the reaction (C_2 axis for the former, multiple elements for the latter), but also in the fact that the roles of the two ethylene molecules are different in the $_\pi 2_s + _\pi 2_a$ cycloaddition. Thus, no symmetry elements relate one ethylene to the other, so we do not mix the reactant orbitals like we do for the $_\pi 2_s + _\pi 2_s$ cycloaddition (Fig. 15.1). The C_2 axis is perpendicular to the page (the line of sight) in the drawings below.

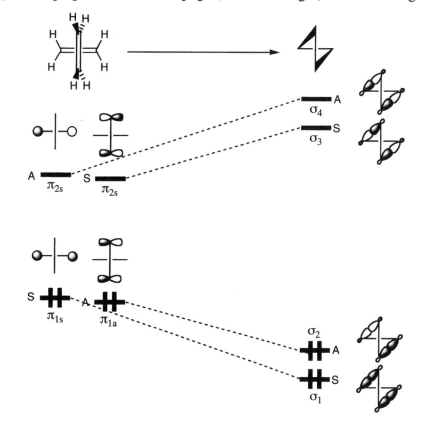

The diagram shows that the ground state of the reactants correlates with the ground state of the product. Therefore, the reaction is allowed.

Based upon this orbital correlation diagram, a state correlation diagram can be generated:

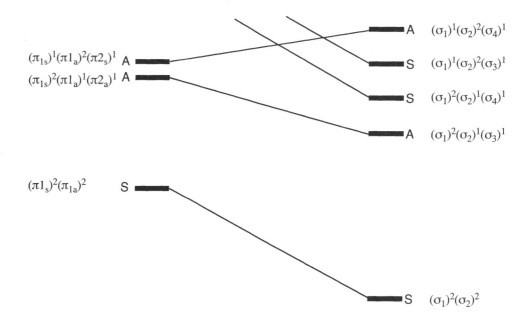

To construct this diagram, one must consider qualitatively the relative energies of the ground and various excited states of the reactants and product. In the diagram above, each state is identified by its electron configuration, with the orbitals labeled as in the orbital diagram. The next step is to determine the symmetry of each state from its electron configuration: a state will be antisymmetric only if it has an odd number of electrons in A orbitals. Finally, the correlated product state for each reactant state is found by starting with the reactant electron configuration and placing each electron into its correlated product orbital. Since the ground states correlate with each other, this reaction is thermally allowed.

GOING DEEPER

Another difference in the $_\pi 2_s + _\pi 2_a$ relative to the $_\pi 2_s + _\pi 2_s$ cycloaddition is that the degeneracy of the π and π^* orbitals of the two ethylenes has been removed, and this has been represented in the orbital mixing diagram. As the ethylenes approach, the antisymmetric π and π^* orbitals will mix, causing the π orbital to drop and the π^* orbital to rise slightly in energy. If the two ethylenes are nearly perpendicular, as shown, the symmetric pair will not mix significantly due to cancellation of overlap. Note that this "pre-mixing" is a second order effect that could easily be ignored, and that other second order effects, such as mixing of the sterically close C–H and π orbitals, have not been included. The point is that the exact appearance of a correlation diagram depends on the assumptions made and the level of theory applied, but we can usually do pretty well in a qualitative sense by ignoring such second order effects, especially if all we want is to see whether the reaction is allowed.

2. The reaction geometries are:

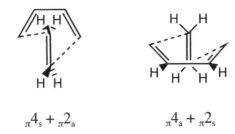

$$_\pi 4_s + {_\pi 2_a} \qquad\qquad {_\pi 4_a} + {_\pi 2_s}$$

For each reaction type, we can consider two different HOMO-LUMO interactions.

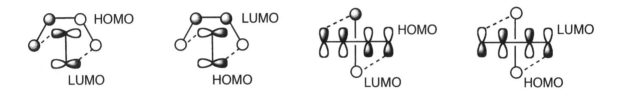

In each of the HOMO-LUMO pairs, interaction is impossible. The HOMO and LUMO are mismatched in symmetry (one in-phase and one out-of-phase interaction), meaning that the reactions are forbidden.

The aromatic transition state method gives the same result. Both reactions, having an odd number of nodes, have Möbius topologies.

With six electrons (4n+2), these reactions are forbidden.

3. The surface area of the combined reactants is greater for the *exo* stereochemistry, leading to increased exposure to the water solvent. Therefore, the more compact *endo* geometry is favored due to the hydrophobic effect.

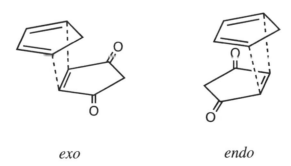

exo   *endo*

4. Only one symmetry element is preserved in this reaction: the mirror plane that bisects both reactants (the plane of the paper as the molecules are drawn below). Since there is no symmetry that forces us to do it, we need not mix the reactant orbitals. The occupation shown is that for allyl cation.

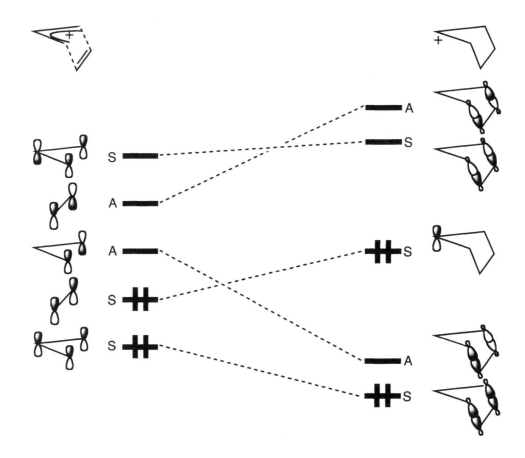

The reaction between allyl cation and ethylene is clearly forbidden, since the orbital correlation leads to an excited state of cyclopentyl cation. Considering the reaction of allyl radical, the additional electron in the allyl NBMO helps, but an excited state of cyclopentyl radical is still obtained. Completely filling the allyl NBMO (allyl anion) gives a correlation with the ground state of cyclopentyl anion, so the reaction of allyl anion and ethylene is allowed.

Note similarities to the [4+2] cycloaddition of butadiene and ethylene (Fig. 15.2). Given this, it is no surprise that we need six electrons for an allowed reaction.

5. The orbital correlation diagram for ozone + ethylene is essentially the same as that for allyl anion + ethylene (exercise 4). The π orbital energies are lower for ozone than for allyl anion, but the symmetry properties are the same. Therefore the reaction is allowed.

6. The C–C bond in the 3-membered ring cleaves, giving a resonance-stabilized zwitterion in which the negative charge is stabilized by the carboxyl groups. This cleavage is a 4-electron electrocyclic reaction, counting the two σ bonding electrons and the N lone pair. The second reaction is a [4 + 2] cycloaddition.

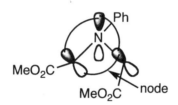

The electrocyclic opening must be conrotatory to be allowed, as shown by an aromatic transition state analysis. An allowed 4-electron reaction should have a Möbius topology, and conrotation gives an odd number of nodes. The *cis,trans* intermediate leads to a trans product following the allowed [4 + 2] cycloaddition.

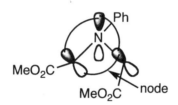

7. In the electrocyclic reaction of hexatriene to cyclohexadiene, the reactant and product each have two mirror planes and a C_2 axis of symmetry. However, not all of the symmetry elements are preserved throughout the reaction, and which ones are preserved depend on the reaction geometry. The conrotatory reaction preserves only the C_2 axis, while the disrotatory reaction preserves only the mirror plane that bisects the molecular plane. Therefore, we should do separate diagrams for these two geometries. Even though the reactant and product orbitals are the same, they have different symmetry labels.

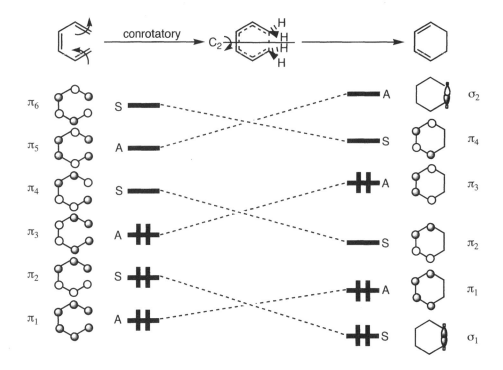

Since the ground state of the reactant correlates to an excited state of the product, the conrotatory reaction is forbidden.

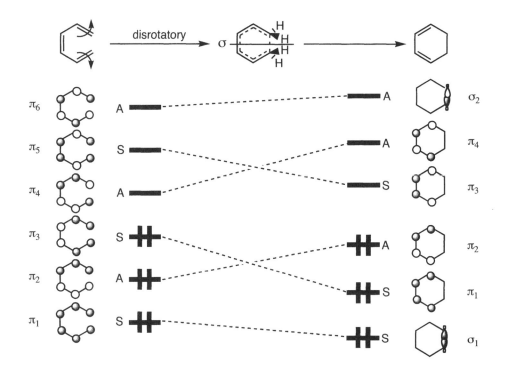

Since the ground state of the reactant correlates to an excited state of the product, the disrotatory reaction is allowed.

8. Based on the orbital correlation diagram in Figure 15.16, the state correlation diagram below can be derived for the conrotatory process. Orbital labels are numbered starting from low to high energy. State symmetries are easily calculated: only states with odd numbers of electrons in A orbitals will be A (so doubly occupied orbitals can be ignored). The reactant state with the electron configuration, $(\pi_1)^2(\pi_2)^1(\pi_3)^1$, is therefore of A symmetry, having one electron in an S orbital (π_2) and one electron in an A orbital (π_3).

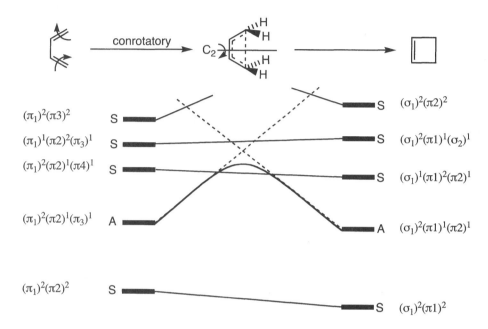

The two ground states are directly correlated, showing that the reaction is thermally allowed. The first excited states correlate to much higher energy states, but since state correlations of the same symmetry cannot cross, this leads to an avoided crossing with a large barrier. The intended correlations are shown as dotted lines, and the actual correlations are shown with solid lines.

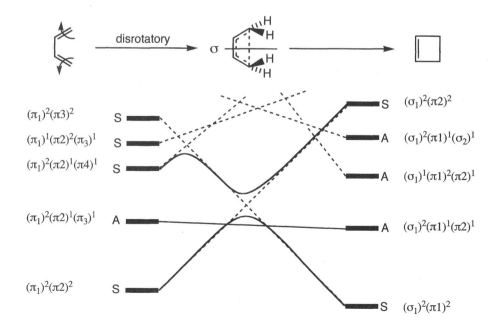

For the disrotatory process, the states are the same, but the conserved symmetry, state symmetries, and correlations are different. The ground states now show intended correlations to highly excited states, and a large barrier on the ground-state surface arises from an avoided crossing.

9. Referring to the orbital correlation diagrams in exercise 7 allows us to construct corresponding state correlation diagrams.

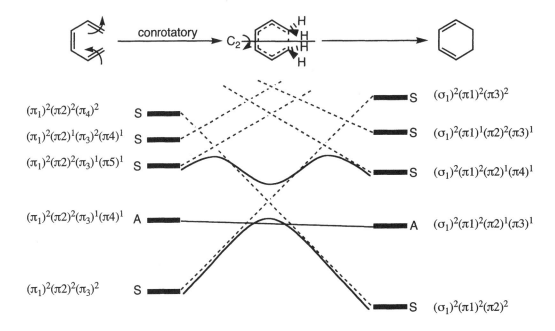

The intended correlations between the ground states and highly excited states result in a large barrier to the conrotatory reaction through an avoided crossing. (Though not relevant to the thermal reaction, note that multiple avoided crossings appear on some excited state surfaces.)

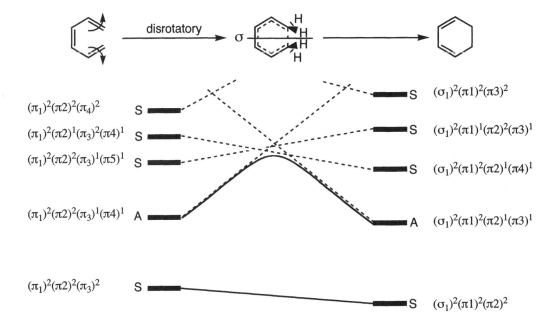

For the disrotatory case, even though the orbital symmetries change, the state symmetries do not for the states shown. But the state correlations do change, and the two ground states are now directly correlated. The absence of an avoided crossing on the ground state surface results in an allowed reaction.

10. The thermal reaction of a cyclobutene is likely to involve an electrocyclic ring opening to a butadiene. In this case, allowed, conrotatory paths can lead to two different products:

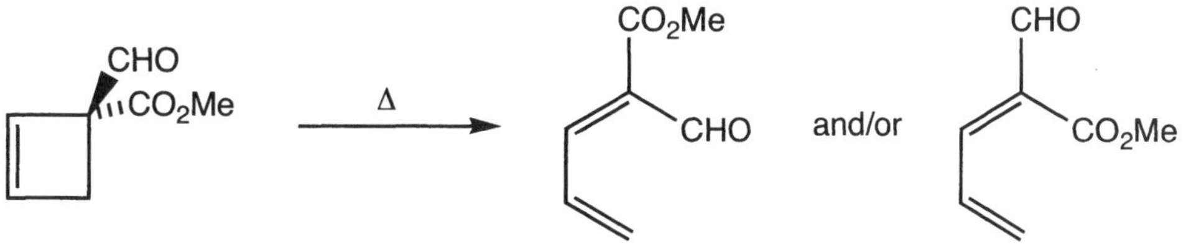

Torquoselectivity in these reactions generally places the more electron-withdrawing substituent in the cis position. In this case, both substituents are electron-withdrawing, but a formyl group is somewhat stronger than a carboxyl group. For example, the pK_a of acetaldehyde is 17, while the pK_a of ethyl acetate is 23. Thus, the first product should dominate. This, clearly, is also the product that more readily leads to the observed product. The second step is also electrocyclic:

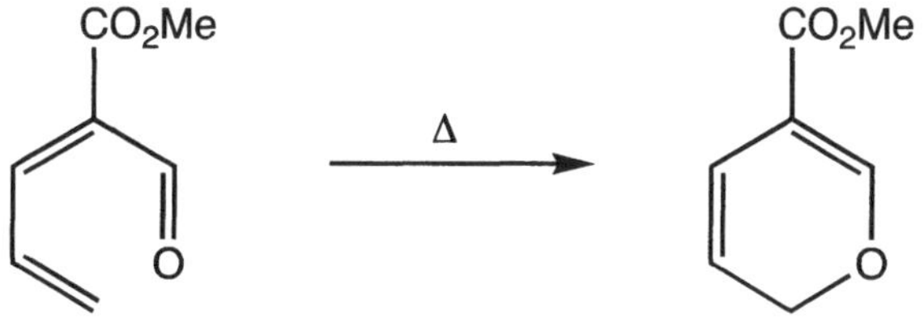

11. Two possible products can be formed through conrotatory pathways:

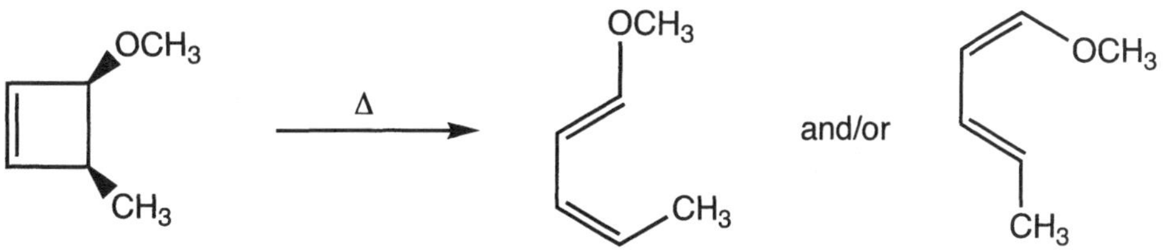

The rule for torquoselectivity in cyclobutene openings is that electron-donating groups like OCH_3 or CH_3 generally prefer placement in the trans position. Both products accomplish this for only one of the substituents. Since the π-donating OCH_3 should exhibit the stronger preference, the first product should predominate.

12. Considering the mechanism first without worrying about stereochemistry might be easiest:

The ring opening of a cyclopropyl cation to an allyl cation is an electrocyclic process. Since it involves two electrons (4n+2, where n = 0), we need a Hückel process, as shown below, in order for it to be allowed. This requires the reaction to be disrotatory. The ring opening of a cyclopropyl cation is very fast, and it is clear that the rate-determining step of the sequence above is the first step. Consider now how the stereochemical placement of the chloride might impact the reaction:

endo-Cl   *exo*-Cl

We can see that in the *endo* chloride, the electrons in the bent (and ready to break) cyclopropyl bond reside on the backside of the C–Cl bond. If the C–Cl cleavage and ring-opening occur simultaneously, the cyclopropyl σ electrons can assist the C–Cl cleavage in a manner similar to a π or lone pair neighboring group. Note that as the disrotatory process begins these two electron move even more into the backside region of the C–Cl bond. However, in the *exo* isomer, these electrons are not well placed for backside assistance, and the C-Cl cleavage therefore has a higher barrier.

13. The most acidic sites in these compounds are adjacent to the cyano group, so we should expect to first obtain cyclopropyl anions:

The electrocyclic opening of a cyclopropyl anion to an allyl anion, being a 4-electron (4n) process, must occur in a conrotatory fashion. Considering what products this leads to gives us a striking difference between the compounds:

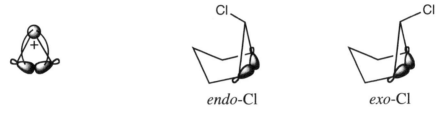

1

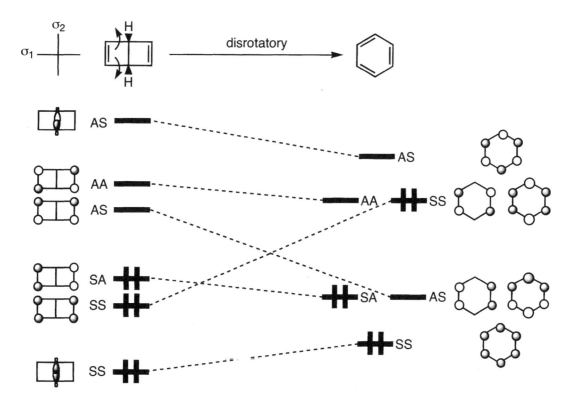

2

The allyl anion that would come from compound **2** would be unreasonably strained, having a *transoid* bond in a 6-membered ring. This explains why no allyl anion is observed from **2**: the opening of the cyclopropyl anion to the reasonable all-*cisoid* allyl anion is a disrotatory, forbidden process.

14. The conversion of Dewar benzene to benzene is disrotatory (the conrotatory process gives a very strained isomer with a *transoid* bond). During this process, two mirror planes are conserved, as well as a C_2 axis. For the orbital correlation diagram, we will use only the mirror planes. (Including the C_2 would not change the result, since the C_2 symmetry is determined by the two mirror planes.)

The ground state of Dewar benzene correlates with an excited state of benzene, so the reaction is forbidden.

15. This is a Cope reaction, a [3,3] sigmatropic shift, for which a chair-like transition state is generally favored. We can explain the formation of either product through a chair-like transition state:

Given that the A-value (energetic preference of equatorial over axial position) is 2.9 kcal/mol for Ph and 1.8 kcal/mol for Me, the observed preference for the product that comes from the equatorial Ph is not surprising.

16. Using Tables 2.4 and 2.7, heats of formation of the reactant/product (1,5-hexadiene) and possible intermediates can be computed:

$$\Delta H_f^\circ = 2\Delta H_f^\circ[C_d-(H)_2] + 2\Delta H_f^\circ[C_d-(H)(C)] + 2\Delta H_f^\circ[C-(C_d)(C)(H)_2]$$

$$= 2(6.26) + 2(8.59) + 2(-4.76) = 20.18 \text{ kcal/mol}$$

$$\Delta H_f^\circ = 2\Delta H_f^\circ[\bullet C-(C)_2(H)] + 4\Delta H_f^\circ[C-(C\bullet)(C)(H)_2]$$

$$= 2(37.45) + 4(-4.95) = 55.10 \text{ kcal/mol}$$

$$\Delta H_f^\circ = 2\Delta H_f^\circ[\bullet C-(H)_2(C_d)] + 2\Delta H_f^\circ[C_d-(C\bullet)(H)] + 2\Delta H_f^\circ[C_d-(H)_2]$$

$$= 2(23.2) + 2(8.59) + 2(6.26) = 76.10 \text{ kcal/mol}$$

The values obtained match those given in the text near Figure 15.24. Note that the stabilization due to allylic resonance requires no correction, since it is reflected in the group increments.

17. The more diffuse charge of K^+ makes the larger cation less capable of ion pairing with the anion. Therefore, the charge on the anion is more free to delocalize and presumably to stabilize the partially forming radical center.

18. The stereochemistry is consistent with the participation of chair-like transition states:

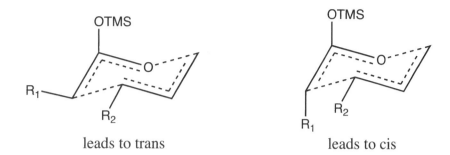

leads to trans   leads to cis

Thus, the enol ether derived from the *Z*-enolate gives the trans product, while the *E* isomer gives the cis product.

19. The racemization mechanism for most sulfoxides is pyramidal inversion at S, and the barrier for this process is near 40 kcal/mol. For allyl sulfoxides, an additional mechanism is possible, involving sequential [2,3] sigmatropic shifts:

The barrier for these shifts is approximately half of that for the inversion process.

20. Considering the LUMO of ethylene and the HOMO of the carbene gives the same result as Figure 15.29: preference for a non-linear approach. In the linear approach, no mixing is possible since the two orbitals are A and S with respect to the horizontal mirror plane. (Any positive overlap is cancelled by an equal negative overlap.) The symmetry is broken in the non-linear approach, allowing good overlap. This FMO analysis thus shows that the linear approach is forbidden and the nonlinear approach is allowed.

linear approach   nonlinear approach

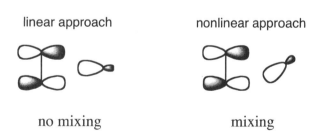

no mixing   mixing

21. The rearrangement of *cis*-divinylcyclopropane is an example of the Cope reaction. It differs from most Cope reactions in that it must proceed through a boat transition state. To prevent the formation of trans double bonds in the product, the vinyl groups must both be oriented under the ring (*endo*), enforcing the boat geometry:

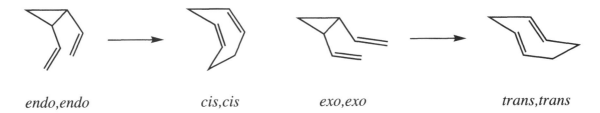

endo,endo   cis,cis   exo,exo   trans,trans

Understanding the necessity for achieving the *endo,endo* conformation, we can easily see that the introduction of methyl groups leads to steric hindrances in the required conformation:

Note that these steric interactions would persist in the transition state, leading to a higher barrier by approximately the energy cost of forming the *endo,endo* conformation – apparently in the range of 5 kcal/mol (assuming the steric cost is same in the reactant and transition state).

22. The product stereochemistry shown is the result of a suprafacial-inversion path for the [1,3] sigmatropic shift. In other words, the migration involves a single face of the allyl π system, and the CXY center undergoes inversion. Using the aromatic transition state method, this transition state has a Möbius topology, and with 4 electrons the reaction is allowed.

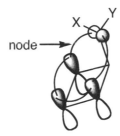

23. Some of these conversions may be hard to see at first. The best procedure is to first identify the bonds that must break and form. Then draw the atomic orbitals involved in the bonds that are breaking in phase (*i.e.*, draw the σ or π bonding orbitals). Next, determine how these orbitals will overlap to form the new bonds. To do this, it may help to redraw the molecule in a geometry that is partially deformed toward the product. Drawing the product with its orbitals may also help. (Models can also be very helpful for visualizing these bonding changes. Not all model sets allow the formation of strained rings, but even a partial model might help.) Once the new interactions are identified, the original bonds can be labeled as

suprafacial or antarafacial, depending on whether the interacting lobes have the same or opposite shading (see Figure 15.9). Note that there may be more than one way to label a reaction, but the generalized orbital symmetry rule should give the same result. For example, a reaction might be classified as $_\sigma2_s + _\sigma2_s$ or $_\sigma2_a + _\sigma2_a$, but in either case it is forbidden.

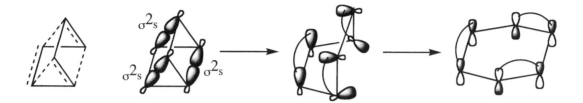

The reactant (prismane) structure at the left shows the three σ bonds that must break and the three π bonds that must form. With three 2-electron suprafacial components, $_\sigma2_s + _\sigma2_s + _\sigma2_s$, this reaction should be allowed. (Note, as described above, that the $_\sigma2_s$ labels depend not on the reactant structure but on the new interactions that form later.) However, this analysis is a simplification of reality, since we are treating the three new π bonds as if they were isolated from each other when indeed they are held in very close, bonding proximity. In other words, this reaction involves more than just a monocyclic array of orbitals. Therefore, the use of the generalized rule in this case is insufficient to provide an answer concerning whether the reaction is allowed or forbidden. Indeed, an orbital correlation diagram gives the opposite conclusion, that the reaction is forbidden. It turns out that the key point from the orbital correlation involves the MOs that are depicted above. Thus, the σ bonding reactant orbital shown correlates with the benzene π MO shown, which happens to be an antibonding orbital. So the ground state reactant correlates with an excited state of the product, and the reaction is forbidden. This is consistent with the remarkable stability of prismane in spite of its high strain. (It decomposes at 90° with an 11 h half life.)

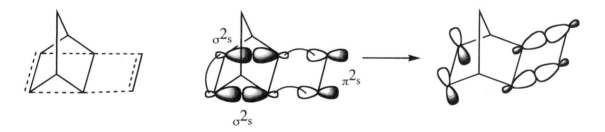

With three 2-electron suprafacial components, $_\sigma2_s + _\sigma2_s + _\pi2_s$, this reaction is allowed. (Note that if we had connected the shaded lobes of the forming π bond, this would become a $_\sigma2_a + _\sigma2_a + _\pi2_s$ reaction, but still allowed.)

Without considering the H atoms that are shown, we might be tempted to just reconnect the large lobes to form the new σ bonds, making this a $_\sigma2_s + _\sigma2_s$ reaction. However rotating the orbitals to accomplish this would place the H atoms in unreasonable positions. Whether we make this mistake or get the proper $_\sigma2_a + _\sigma2_a$ designation for this reaction, we still get the same result from the rule: forbidden.

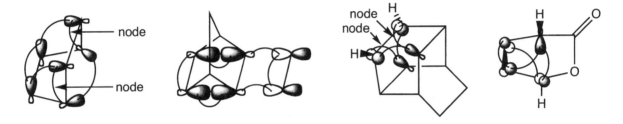

According to the generalized rule, this $_\pi2_a + _\sigma2_a$ reaction is forbidden.

24. We can use one drawing from each answer to use the aromatic transition state method.

It turns out that in each case there are either zero or two nodes present, so these all have Hückel topology. The first two, having six electrons, should be allowed, while the last two, having four electrons, should be forbidden. As noted in exercise 23, however, the conversion of prismane to benzene has more orbital interactions than the ones represented in the cyclic array connected by arcs. So the aromatic transition state method has the same shortcomings in this case and gives the same false answer.

25. Since these are reactions that actually happen, we should try to find mechanisms in which all steps are allowed. In the first reaction, the absence of any conjugation or rings limits the possible reactions. An ene reaction, followed by a [1,3] shift, produces the product:

Note that this [1,3] shift, called the vinylcyclopropane rearrangement, is much more favorable because of the ring strain relief. In order to be allowed, this [1,3] shift would have to be either suprafacial-inversion or antarafacial-retention. Mechanistic studies on vinylcyclopropane rearrangements have shown that both allowed and forbidden paths are followed, suggesting a biradical mechanism (see Chapter 11, exercise 36). The lack of stereochemistry in this example would preclude any conclusions about the mechanism in this case.

The next reaction looks like a [2+2] retrocycloaddition, but this reaction would not be allowed. The presence of a cyclohexene ring suggests that we might try a retro-Diels-Alder:

Only one additional step is required to get to the observed product, a Cope reaction. The Cope is difficult to recognize in a setting like this, but one can both recognize the possibility and work out the product by focusing on the required functionality, a 1,5-hexadiene. Numbering carbons is often helpful in a case like this. Another technique used here is to draw the product first in the same geometry, even though a badly distorted structure is obtained. From here it is easier to recognize that the desired product is indeed formed. In order for this Cope reaction to be viable, carbons 1 and 6 must be able to approach each other in the reactant, and this is clearly possible.

In the last reaction, the fused five-membered rings of the product are apparent in the reactant, and the extrusion of ethylene should be possible in a retro-Diels-Alder reaction if one of the double bonds can be moved into the norbornane six-membered ring. This can be accomplished through a very common, allowed reaction: a [1,5] sigmatropic H shift, a reaction that is especially favorable in cyclopentadienes.

We are not done yet, but we can get to the final product with two more [1,5] shifts.

These H-shift isomers might be expected to coexist in the product mixture.

26. Presuming that both reactions are suprafacial with respect to the conjugated ring, we only need to determine whether the reaction goes with retention or inversion at the migrating carbon. If there is retention, the a and b groups will switch places as the Cab group rotates to form the new bond. With inversion at C, the a group will always remain in the *endo* position.

The [1,5] shift involves six electrons, so it will be allowed with a Hückel topology, corresponding to retention. The a and b groups switch positions.

This [1,4] shift involves four electrons, so it will be allowed with a Möbius topology, corresponding to inversion. The a and b groups do not switch positions.

27. To predict regiochemistry, we need to first identify the more important HOMO-LUMO interaction, and then determine which regiochemistry produces higher overlap. The first example is a normal electron-demand Diels-Alder reaction, with an electron-rich diene and an electron-poor dienophile. The important frontier interaction involves the HOMO of the diene and the LUMO of the dienophile. Both of these orbitals have their larger coefficients away from the substituent (see Figure 15.12), so the best overlap in the Diels-Alder reaction is obtained in the formation of the 3,4-regioisomer shown. The *endo* preference leads to the stereochemistry shown.

The second example involves a similar diene and the same dienophile. The coefficients are reversed in this diene, leading to a preference for the 1,4-product.

The third example is an inverse electron-demand Diels-Alder, with an electron-poor diene and an electron-rich dienophile. The important frontier MO interaction is between the LUMO of the diene and the HOMO of the dienophile. The amine group should end up near the O. (Carbonyl groups are known to participate as part of the diene in Diels-Alder reactions, but these reactions are less common.)

28. In this reaction, ethyl acrylate would be considered a "good" dienophile, meaning that the reaction is accelerated by the electron withdrawing carboxyl group. We can understand this as arising from a reduction of the energy gap for the dominant frontier orbital interaction between the HOMO of the diene and the LUMO of the dienophile. A Lewis acid like $AlCl_3$ will coordinate to the carboxyl group, further lowering the LUMO of the dienophile and accelerating the reaction.

29. In order to achieve constructive interactions between the HOMO and LUMO of conjugated ethylene units, the top of one end p orbital would need to overlap with the bottom of the other. This requires conrotation:

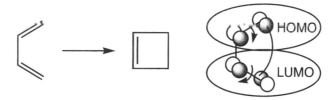

In a similar interaction between conjugated butadiene and ethylene units, disrotation is required. The same result is obtained for either choice of HOMO-LUMO pairings.

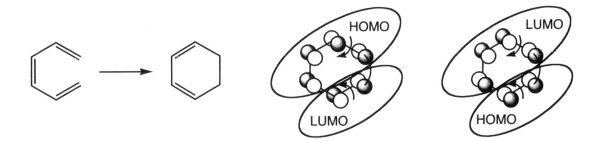

30. This method is slightly simpler than the one in exercise 29. The theoretical justification for it is not as clear, but this does not mean it should not be used – after all, it does give the same answer. The HOMOs of butadiene and hexatriene predict conrotation and disrotation, respectively:

31. The thermally allowed reactions for these reactants are ene reactions:

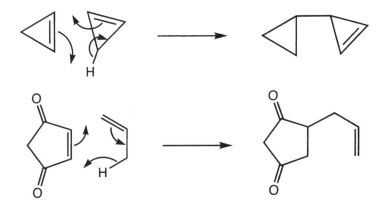

32. Actually, two pericyclic reactions are involved: a Claisen reaction and a decarboxylation reaction (though the latter is better classified as pseudopericyclic; see exercise 37). Note that the equilibrium for the first tautomerization step favors the enol form, due to the formation of a hydrogen bond that forms a six-membered ring.

33. Knowing that cyclopentadiene is a good Diels-Alder diene, we should look for that possibility. A Diels-Alder reaction followed by a [1,3] alkyl shift should produce the observed product. Note that the phenyl groups would provide stabilization to a biradical intermediate for the [1,3] shift, suggesting that this reaction might be stepwise rather than concerted.

34. The reaction shown is the reverse of the ene reaction in the first part of exercise 25, so it indeed is a retro-ene reaction. The cis stereochemistry of the product arises from the six-membered transition state. Only the *s-cis* conformation, leading to the cis product, will be reactive, since the *s-trans* conformation places the alkene and methyl H groups too far apart. This is shown by the Newman projections of the reactant, as viewed along the bond whose stereochemistry is in question.

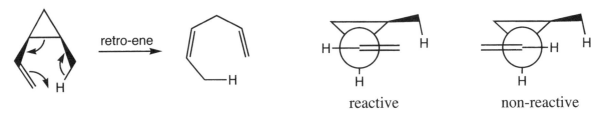

35. The orbital correlation for the transfer of two H atoms from ethane to ethylene is based upon the mirror plane that bisects the molecules, a horizontal plane in the molecular orientation shown.

The ground states clearly correlate, so the reaction is allowed.

FMO analysis can be done in two ways, shown below, and both show that the reaction is allowed. The HOMO of ethane is the same symmetry (A) as the LUMO of ethylene, and the HOMO of ethylene is the same symmetry (S) as the LUMO of ethane.

HOMO LUMO   LUMO HOMO

Using the aromatic transition state method, we see that the reaction has a Hückel topology (zero nodes) and six electrons and is therefore allowed.

Finally, the generalized selection rule agrees that the reaction is allowed as a $_\sigma 2_s + _\sigma 2_s + _\pi 2_s$ reaction.

Since the model reaction is allowed, and since the example shown is especially favorable due to the formation of an aromatic naphthalene molecule, it seems that the reaction should occur readily. (It is indeed observed.)

GOING DEEPER

The orbital correlation diagram above is remarkably simple and might make us feel like we have done something wrong. Indeed, it appears that this reaction will be allowed regardless of the number of electrons occupying the orbitals! If we add two electrons, so that we have eight (4n), the reaction will still be allowed. Granted, the dianion of ethylene is not a very realistic reactant, but the same sort of diagram should apply to any identity reaction that has a bisecting symmetry element. It so happens that such reactions, possible for both atom-transfer and sigmatropic reactions (*e.g.*, the Cope reaction), have 4n+2 electrons for the neutral cases. However, for some of the larger reactions, doubly charged species might become reasonable. This intriguing result from the correlation diagram does not come out from any of our simplified methods: FMO, aromatic transition state, or the generalized selection rule.

36. The conrotatory reaction, $_\sigma 2_a + _\omega 2_s$, is allowed, while the disrotatory reaction, $_\sigma 2_s + _\omega 2_s$, is forbidden. The expected product is the *E,Z*-dimethylallyl anion.

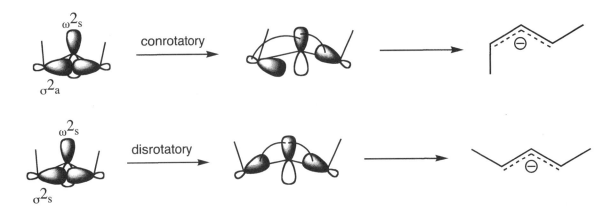

37. Though we could draw arrows in the same manner as for hydrocarbon sigmatropic rearrangements, involving only the σ and π bonds, this is not correct in this case. The C=C p orbitals can overlap much better with the O lone pairs than with the O p orbital involved in the C=O bond. In other words, the presence of the O lone pair allows the transition state to have good overlap for all forming and breaking bonds with less strain.

A pericyclic reaction is one that "involves a transition state with a cyclic array of atoms and an associated cyclic array of interacting orbitals." Even though there are cyclic arrays of orbitals, these reactions are not pericyclic because the cyclic arrays are not overlapping everywhere. The two AOs at each O, the lone pairs and the AOs involved in π or σ bonding, are orthogonal to each other. So "interacting" is the key piece that is missing.

CHAPTER

16

Photochemistry

S O L U T I O N S   T O   E X E R C I S E S

1. Consider the lifetime of A* and quantum yield for product formation both in the absence of quencher (a) and presence of quencher (Q).

$$\frac{1}{\tau_a} = k_1 + k_{rxn} \quad \text{and} \quad \Phi_{rxn}(a) = \frac{k_{rxn}}{k_1 + k_{rxn}} = \tau_a k_{rxn}$$

$$\frac{1}{\tau_Q} = k_1 + k_{rxn} + k_q[Q] \quad \text{and} \quad \Phi_{rxn}(Q) = \frac{k_{rxn}}{k_1 + k_{rxn} + k_q[Q]} = \tau_Q k_{rxn}$$

The relative quantum yield is then

$$\Phi_{rxn}(rel) = \frac{\Phi_{rxn}(a)}{\Phi_{rxn}(Q)} = \frac{\tau_a k_{rxn}}{\tau_Q k_{rxn}} = \frac{\tau_a}{\tau_Q} = \frac{k_1 + k_{rxn} + k_q[Q]}{k_1 + k_{rxn}} = 1 + k_q \tau_a[Q]$$

So a plot of $\Phi_{rxn}(rel)$ vs. [Q] should be linear, with a slope of $k_q \tau_a$ and an intercept of 1. If we assume a value for k_q, we can obtain τ_a. If desired, τ_Q can also be calculated for any [Q].

2. We do not know what D and A are, so we do not know what their MOs look like (non-bonding, π system, etc.) But we can assume for simplicity that D and A each have one atom with a hybrid (sp³-like) orbital. For D this orbital is filled, and for A it is empty. For the best overlap, we point them at each other.

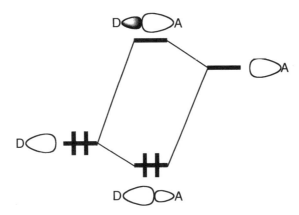

In the first excited state each of the D–A MOs has one electron. Since the bonding orbital is mostly D and the antibonding orbital is mostly A, the excited state has essentially one electron associated with D and one with A. In other words, the excitation is roughly equivalent to an electron transfer from D to A, producing D^+–A^-.

3. As shown in exercise 2, the excited state is strongly dipolar, so it should be strongly solvated by polar solvents. In other words, with its stronger dipole the excited state is stabilized more than the ground state in polar solvents. This results in a lower energy, longer wavelength transition.

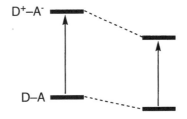

GOING DEEPER

A little deeper thought about this might lead you to this question: If the absorption is a "vertical" transition as the Franck-Condon principle says it must be, how can the solvent molecules respond to the excited state charges in order to stabilize them? Indeed the absorption is much faster than any movement of the solvent molecules. Therefore, the solvation of the forming excited state is only that offered by the arrangement of solvent molecules around the ground state. But there are similar partial charges in the ground state, only smaller. (Referring to the MO diagram of exercise 2, we see that occupation of the bonding orbital with two electrons does serve to give some of the electron density to A.) Thus, the optimum solvation of the ground and excited states should involve similar arrangements of solvent molecules. In other words, the partial charges in the ground state serve to order the solvent in a way that is even more favorable for the excited state with its even greater partial charges.

4. The emission observed after the 1 ns delay is the fluorescence of isolated pyrene molecules. The band structure and small Stokes shift reflect the rigid nature of the molecule and the resulting small geometry change upon excitation. The longer 100 ns delay allows the formation of the excimer, which fluoresces at longer wavelengths with a broader band (see exercise 6).

5. The long-wavelength fluorescence must be due to excimer fluorescence. Formation of an excimer would be favored when two naphthalene units are covalently attached with an appropriate linker. The attachment shown apparently better promotes excimer formation compared to other regioisomers. The isomer shown should allow a relatively unstrained face-to-face interaction.

6. Charge transfer complexes and exciplexes are held together with fairly weak forces. Therefore, the geometries (intermolecular distance, relative orientation) within these complexes can vary quite significantly – certainly more than the variation expected within a single molecule. The larger range of geometries leads to a larger range of wavelengths of absorption and emission. In effect, the band represents the sum of the bands from the various geometries that are absorbing or fluorescing. Any vibrational structure of one geometry is averaged with that from many others, such that the band becomes broad and devoid of vibrational structure.

7. FRET is used to obtain the distance between two chromophores in a biomolecule or complex. In this case, the $3\cos^2\theta - 1$ term is a correction for the relative orientation of the two chromophores. If the biomolecule or complex is in solution and therefore tumbling, this does not change the *relative* orientation of the chromophores. Therefore, tumbling causes no averaging of this orientation term. In contrast, the NMR orientation term refers to the relative orientations of the nuclear dipole and the magnetic field, so tumbling does cause averaging in this case. Another important point is that the time scale of energy transfer is fast (typically nanoseconds) and generally faster than the tumbling of a large biomolecule.

GOING DEEPER

Having made these arguments for why the orientation term is needed, we should also state that analysis of orientational contributions to FRET results can be quite complicated. Both static distributions of relative orientations of the two chromophores and dynamic motions of the chromophores can affect results. With careful measurements and interpretation, these complexities can sometimes be sorted out to provide additional information about the biomolecule structure and dynamics.

8. We can use Eq. 16.18 to analyze the photostationary state:

$$\frac{[T]}{[C]} = \frac{\varepsilon_C \Phi_C}{\varepsilon_T \Phi_T}$$

The quantum yields given in Eq. 16.19 should promote only a small preference for the cis isomer in the photostationary state. Thus, the preference of cis must result from a favorable ratio of extinction coefficients. We can compute the ratio necessary to achieve the observed result:

$$\frac{7\%}{93\%} = \frac{\varepsilon_C (0.35)}{\varepsilon_T (0.50)}$$

$$\frac{\varepsilon_C}{\varepsilon_T} = 0.11$$

This shows that the extinction coefficients and quantum yield both favor the cis isomer, but the extinction coefficients are the more important factor.

9. Comparing single and simultaneous double rotation in a zwitterionic state, we can identify two effects that would favor the former:
   (1) allylic delocalization of one of the charges
   (2) less separation of the opposite charges (a Coulombic effect)

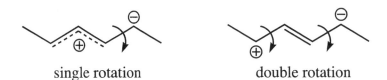

single rotation   double rotation

In a biradical-like state without zwitterionic character, the first of these would still apply, but the second would not. Therefore, double rotation might be better able to compete in the triplet excited state.

Calculations suggest an alternative explanation. The triplet has been calculated to be best described as an allylic species with relatively free rotation at one end, allowing for single rotation. However, end-to-end isomerization, as shown below, can occur, leading to double rotation. By this mechanism, double rotation occurs sequentially instead of simultaneously. Presumably, the shorter lifetime of the singlet biradical would help to limit this end-to-end isomerization and prevent double rotation.

10. In the cis isomer, there is significant conjugation between the alkene and carbonyl groups, and this is reflected in the red-shifted λ_{max} value. However, in the trans isomer, this conjugation is broken by significant twisting about the (C=C)–(C=O) bond. The λ_{max} value is near that for acetone, suggesting that the dihedral angle is near 90°. Molecular mechanics calculations agree with this observation. (Try the calculation yourself!)

11. The dominant chair conformations of **A** and **B** are the ones in which the t-Bu group is equatorial. The propyl group is therefore equatorial in **A** and axial in **B**. The equatorial orientation allows the roughly equatorial O to easily reach the γ-H, but the axial orientation does not. Therefore, Norrish II chemistry is observed for **A** but not for **B**. Likely products are shown for each isomer.

12. We can rationalize the pH effect by noting that the last step of Eq. 16.67 requires H⁺. At higher pH, the concentration of H⁺ is lower (by definition). (This analysis implies that this last step is rate determining.)

13. The added methyl group makes the radical formed by H-abstraction more stable, facilitating the process. A similar effect was noted in the text for the Norrish II reaction.

14. The following mechanism starts with the formation of an excited state that undergoes a
    tautomerization, probably through a solvent-assisted proton transfer. The tautomer could be
    formed in either the excited or ground state; the excited state is drawn here (as in the
    literature), such that plenty of energy should be available for the formation of the
    cyclopropanone. Then follows nucleophilic opening of the cyclopropanone and several acid-
    base steps. Note that other mechanisms can be drawn.

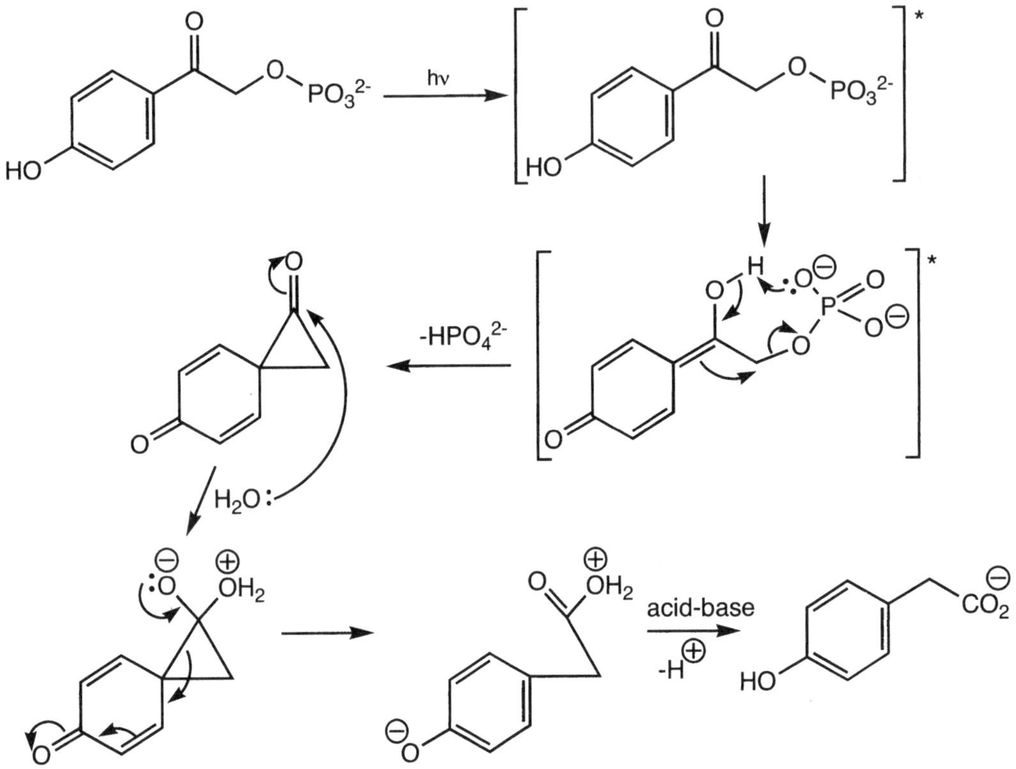

15. The following three-step mechanism accounts for the product formed. Note that only the first
    step is photochemical.

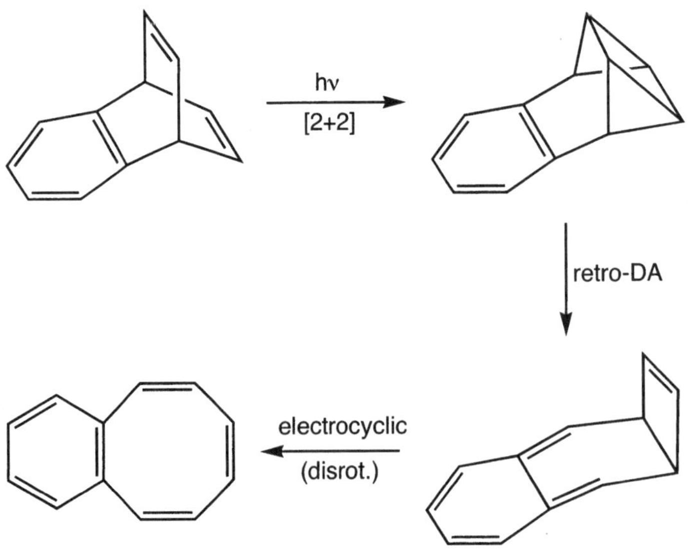

16. This is a di-π-methane reaction, where one of the phenyl groups acts as one of the two π groups:

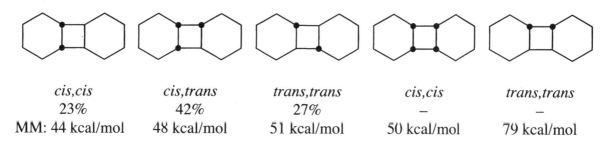

17. The two reactions are done very similarly, having only different sensitizers. Though acetophenone is successful as a sensitizer, promoting the dimerization of norbornene, benzophenone is not successful, instead participating in a Paterno-Buchi reaction. Table 16.5 shows that the two sensitizers have somewhat different triplet energies: 73.6 kcal/mol for acetophenone and 68.5 kcal/mol for benzophenone. So a reasonable explanation is that the triplet energy of norbornane is between these values, such that sufficient energy is available only from acetophenone.

18. The relative yields do not seem to reflect the product thermodynamic stabilities. A cis ring fusion involving a small ring is generally preferred over a trans ring fusion, so the order of product stability should be *cis,cis*, then *cis,trans*, then *trans,trans*. A quick molecular mechanics calculation confirms this thermodynamic ordering, with the total energies shown below. However, the order in terms of yield is different, with the *cis,cis* isomer having the lowest yield of the three observed isomers. Note that two of the five possible diastereomers (the *cis,cis* and *trans,trans* isomers shown at the right) are not listed as products.

cis,cis	*cis,trans*	*trans,trans*	*cis,cis*	*trans,trans*
23%	42%	27%	–	–
MM: 44 kcal/mol	48 kcal/mol	51 kcal/mol	50 kcal/mol	79 kcal/mol

---

GOING DEEPER

If we think about the dimerization reaction, we can make more headway at understanding the product ratios. Since this is a triplet-sensitized reaction, the reaction might be expected to proceed through a biradical intermediate. Initial photochemical isomerization to *trans*-cyclohexene is actually proposed in the literature, followed by thermal reaction between the cis and trans isomers (either concerted or stepwise). Two diastereomers of the intermediate biradical are possible:

We might presume that these isomers would be close in energy and that little selectivity would be encountered, such that each would be responsible for approximately half of the product mixture. The left isomer can close to give all three of the observed products, but the right isomer can give only the third product and the two products that are not observed. In fact, we can fairly well model the observed product percentages with just two postulates: the one already stated (non-selective biradical formation) and that ring closure is selective for an *anti* disposition of H's at the second bond. (We also presume the two *anti* closures to be equally good. This and the second postulate make reasonable sense, considering the free rotation between the rings and the near planarity of the radical centers. Try the different closures with a model!)

Predicted:	25%	25%	50%	0%	0%
Observed:	23%	27%	42%	0%	0%

An interesting issue that arises from the molecular mechanics calculations is the fact that four of the five isomers are within 7 kcal/mol of each other, while the fifth is 28 kcal/mol higher in energy than all the rest! Why should this isomer be so different? The drawings above do not make this obvious, but this isomer is indeed considerably more strained than the others. It has a planar cyclobutane ring that is unable to pucker without applying additional strain to the already over-twisted cyclohexane chairs. In contrast, the other *trans,trans* isomer, though also rigid, is trapped in a more favorable conformation with respect to both the puckered cyclobutane and chair cyclohexane rings. In a sense, the rings work together in this isomer, rather than against each other in the high-energy isomer. Perspective views of the two *trans,trans* isomers show these features to some extent; models show the contrast very clearly.

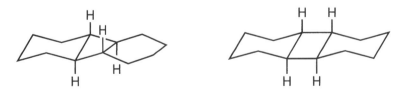

MM2:   51 kcal/mol   79 kcal/mol

19. In the smaller rings, the excitation energy presumably goes toward formation of a biradical (similar to the one in exercise 18) followed by ring closure. In the larger rings, perhaps starting with cyclohexene, the excitation energy can go toward isomerization to a *trans*-cycloalkene. Even in the ground state, *trans*-cyclohexene and *trans*-cycloheptene are energetic enough to react with a cis molecule to produce the analogous biradical. However, *trans*-cyclooctene is much less strained and is not energetic enough to react further. We could call the conformational process in cyclooctene a loose bolt effect in that is serves to dissipate most of the excitation energy and prevent dimerization. In contrast, the isomerization process in cyclohexene or cycloheptene stores some of the energy for later use – like setting a mousetrap!

20. Rotation about the C–C bond shown in the biradical intermediate followed by cleavage to the reactants will lead to formation of both cis and trans olefin isomers. The result also suggests that this rotation must be comparable to or faster than both cleavage and closure of the biradical.

21. The more stable biradical gives the major product.

22. The biradical with a new C–O bond is more stable than the one with a new C–C bond. Referring to Table 2.2B, C–O and C–C bonds are approximately equal in strength, on average, but the C radical originating from the carbonyl C is stabilized by overlap with an O lone pair (2c,3e⁻ bonding). For any aldehyde or ketone other than formaldehyde, the C radical is also stabilized by alkyl or aryl substitution.

23. The product arises from H-abstraction in the biradical, rather than coupling that forms the usual Paterno-Buchi product.

24. Considering the bonding changes that occur in Eq. 16.78 and the observation that chemiluminescence is generally associated with cleavage of O–O bonds, we can envision a general sequence: removal of the hydrazine protons, replacement of N_2 with O_2, and cleavage of the O–O bond to give an excited state. Some of the details are less clear, but here's one way to draw it:

25. The shorter lifetime in benzene (30 μs) relative to acetone (50 μs) reflects the higher C–H stretch for benzene, due to its sp^2-hybridized carbon atoms. Compounds with higher energy vibrations can more efficiently accept energy from singlet oxygen as it goes to its ground triplet state.

26. Cell membranes, being lipid-based, have much less water than the cytosol. Since the lifetime of singlet oxygen is very short in water, the observed trend follows expectations.

27. To explain the lower pK_a values in the excited states, we could either show that the excited acid is stronger or show that the excited base is weaker. Considering the naphthoxide base, the ground state has two electrons in the HOMO, essentially an O lone pair, while in the excited state, one of these electrons is promoted to a π^* orbital. Since the π^* orbital is delocalized and not centered on O, excitation serves to delocalize the negative charge. With less concentrated charge on the O, the excited naphthoxide is a weaker base.

The same argument explains why naphthylamine is a weaker base in the excited state.

28. One difference between azo compounds and olefins is that in azo compounds, the absorption is n,π* in character as opposed to π,π*. The two N lone pairs significantly overlap in a cis azo compound, leading to a filled-filled interaction that produces a higher energy n orbital (the out-of-phase combination). With the n orbital at higher energy, the n,π* transition requires less energy (occurs at longer wavelength).

29. a. Mechanism 2:

exciplex

Mechanism 3:

b. Both azoalkanes should, upon photolysis or thermolysis, produce the biradical intermediate from mechanism 2. Thus, if irradiation of the azoalkanes gives the same products as the *m*-xylene + cyclopentene photolysis, then mechanism 2 would be supported as a possibility. Conversely, if different products are obtained, mechanism 2 might be eliminated from consideration. No direct conclusions could be made concerning mechanisms 1 or 3.

30. The reaction begins with a Norrish II-like H-abstraction, but with the intervening aromatic ring a diene is formed instead of a biradical. The reactive, non-aromatic diene then reacts with diethyl fumarate in a Diels-Alder reaction.

31. Exchange of an unpaired electron in a paramagnetic compound with an electron of the opposite spin in a singlet state will produce a triplet state. Likewise, a triplet is converted to a singlet (the reverse direction below).

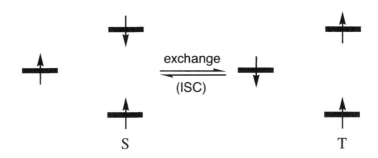

32. A simple MO diagram can be constructed from unhybridized O atoms, where the internuclear axis is the z axis:

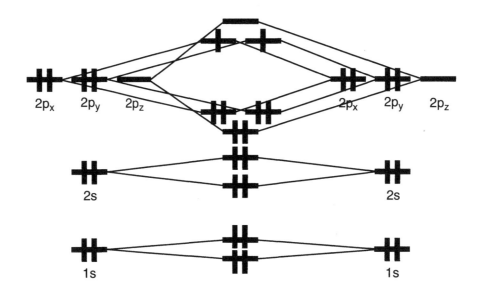

Since each O atom contributes 8 electrons, O_2 has 16 electrons, the last two of which are placed in a degenerate pair of π^*orbitals. Since these orbitals have zero overlap, the additional exchange interaction (K) in the triplet state causes it to be lower in energy. (This is often cited as Hund's Rule, which, in the strictest sense, applies only to atoms. One must take care in applying Hund's Rule to molecules, being sure that the MOs in question are truly degenerate – as they are in O_2.)

33. The important point is that other processes, such as fluorescence and internal conversion, can compete with the thermally activated reaction on S_1. At higher temperatures, the reaction will become faster, but the competing processes will not be dependent on temperature (at least not in the same way). Therefore the likelihood (quantum yield) of the photoreaction is temperature dependent.

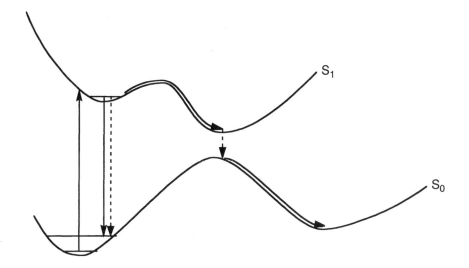

34. If the two triplets have opposite-spin electrons, then electron exchange gives two singlets:

35. The energy of a 185 nm photon, 155 kcal/mol, is enough to easily break even strong σ bonds. As shown, all four products can be formed from the same biradical that results from cleavage of the C_2–C_3 bond. Cleavage of the 1,4-biradical gives ethylene and acetylene. Internal H-abstraction gives a carbene that leads to the other two products through insertion of the carbene into two different C–H bonds. Note that butadiene could also be formed directly by cleavage of the cyclobutene C_3–C_4 bond (electrocyclic ring opening).

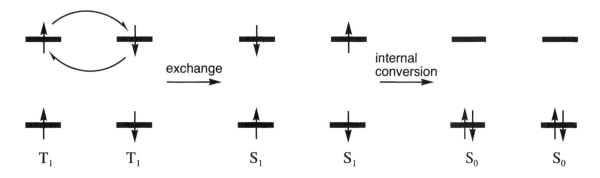

36. Both fulvene and benzvalene could be formed from the same "allyl + p" biradical, the former by C–C cleavage and C–H insertion and the latter by closure.

In the triplet state, the allyl + p biradical could first close to form a C–C bond to the central allyl C, giving another biradical. Cleavage and closure of this 1,4-biradical would produce Dewar benzene and prismane, respectively.

37. Cross-over experiments would serve to distinguish between concerted and stepwise mechanisms. The observation of cross-over products would be inconsistent with a concerted mechanism and would strongly support a stepwise process. The absence of cross-over products would be suggestive of a concerted mechanism, but would not rule out the possibility of a stepwise process with a strong cage effect.

38. A trans-to-cis isomerization places the γ-H close to the carbonyl, such that an abstraction related to a Norrish II reaction can effect the deconjugation:

Note that a concerted 1,3-H sigmatropic shift, not involving the carbonyl directly, could also explain the reaction. Such a reaction may be photochemically allowed, and many examples of photochemical 1,3-H shifts are known, at least from $^1(\pi,\pi^*)$ states. Experiments suggest that the present case goes through n,π* states and that the enol is an intermediate. (In deuterated solvents, the product ketone is formed with D at the α position.)

39. The fact that this reaction only occurs with cyclohexenes and cycloheptenes suggests that isomerization of the alkene from cis to trans occurs. The energy obtained through the sensitizer must be enough to produce either the singlet or triplet excited state of the cyclohexene, resulting in isomerization to the ground-state *trans*-cyclohexene. The very high strain energy makes this alkene very reactive, such that it can be protonated by methanol. The formation of a carbenium ion is consistent with the observed Markovnikov regiochemistry. The photochemistry serves only to produce the strained trans isomer, and everything after that occurs on the ground state surface.

40. The nitro group is strongly electron-withdrawing and the phenyl groups can stabilize either charge. Therefore, the excited state is likely to have significant broken-bond character as depicted by the second structure below. This leads to a simple hydrolysis mechanism.

This mechanism could be tested by running the reaction in $H_2^{18}O$. If the ^{18}O label appears in the triphenylmethanol but not in the 3-nitrophenol, the mechanism above would be supported. An interesting experiment that was reported was the photolysis of the *para* isomer. This isomer is much more reactive in the ground state, but appears to show little or no acceleration with irradiation. The authors noted that the electron distribution, though better for heterolysis in the ground state for the *para* isomer, is better for heterolysis in the excited state for the *meta* isomer (Zimmerman, H. E.; Somasekhara, S. *J. Am. Chem. Soc.* **1963**, *85*, 922).

41. An anti-Stokes shift requires absorption by a molecule that is already in an excited vibrational state. Under normal circumstances, almost all molecules are in the ground vibrational state, so observations of anti-Stokes shifts are rare. Anomalous fluorescence is fluorescence that occurs from a state higher than S_1. This is also rare, since internal conversion among excited states is generally very efficient.

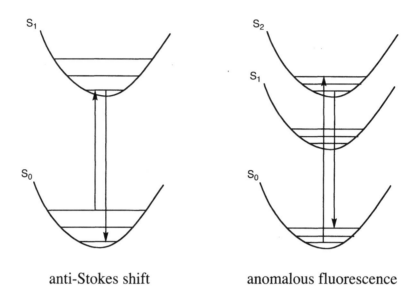

anti-Stokes shift   anomalous fluorescence

42. The Franck-Condon factor is important, because absorption and fluorescence occur very fast, essentially instantaneously relative to the motions of solvent molecules. The excited state, with its partial charges, should be more strongly and specifically solvated, but it is formed with the solvent molecules in non-optimum positions and orientations. After absorption, the excited state energy drops as the solvent responds to the new charges. We can show this on a Jablonski diagram by taking the geometry coordinate as a solvation coordinate, such that movement along the x axis corresponds to movements of the ensemble of solvent molecules (*i.e.*, many degrees of freedom projected onto one axis). On the solvation axis, the energy of the charge-separated excited state should vary more steeply than the ground state. As shown, as long as the optimum arrangements of solvent molecules is different, as argued above, a large Stokes shift is expected.

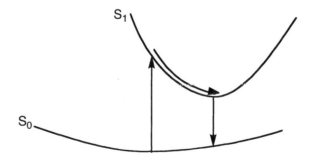

43. a. The presence of the benzene π orbitals lowers the energy of the π,π* transition and causes some mixing of the n,π* and π,π* states. Therefore, the n,π* transition in benzophenone, with its partial π,π* character, is more allowed, giving it a higher extinction coefficient.

   b. A single absorption event can lead to both product formation and product fluorescence, so summing the three quantum yields leads to double accounting for some excited molecules.

   c. The highest energy electrons in a 1,2-diazene are the non-bonding electrons, the N lone pairs (shown). The lowest energy unfilled orbital is the π* orbital for the N=N bond. So the S_0 to S_1 transition is classified as n,π*. Since the lone pairs reside in the nodal plane of the π* orbital, the transition is spatially forbidden.

   d. If quantum yields can be measured for all of the other primary processes, such as fluorescence, intersystem crossing, and product formation, then the quantum yield of internal conversion can be estimated as 1-(sum of the other quantum yields).

44. a. Spin-forbidden, spatially allowed (π,π*), absorption.
    b. Spin-allowed, spatially forbidden (n,π*), internal conversion.
    c. Spin-forbidden, spatially forbidden (n,π*), phosphorescence.

45. 1,3-Pentadiene clearly quenches the triplet state of benzophenone, and in so doing, its T_1 state is presumably formed. The fact that the T_1 state is not detectable suggests that its lifetime is much shorter than that of the benzophenone T_1 state. The pentadiene T_1 state is π,π* in character, so intersystem crossing to S_0 is spatially allowed. Rotation about the C–C π bonds can also serve to dissipate the energy through the loose-bolt effect.

46. In this dimerization, only one of the two molecules must absorb light. If both did, the quantum yield for anthracene disappearance could not exceed 1. Formation of an excimer that leads to product seems likely.

Anthracene $\xrightarrow{\text{hv}}$ Anthracene* $\xrightarrow{\text{Anthracene}}$ (Anthracene)$_2$* $\longrightarrow$ Dimer

excimer

47. Direct absorption to the product S_1 state violates the Franck-Condon principle. A "vertical" transition, allowed by Franck-Condon, is one that goes with no change of geometry.

48. a.

cyclobutanol

enantiomer (racemization)

Type II elimination products

b. The racemization, with a quantum yield of 0.78, can clearly produce the racemate in greater than 25% yield, though this would rarely be our goal. The type II elimination products can also be generated in high yield, much more than 25%, because the racemic starting material is just as photoreactive. With enough light, the starting material can be completely consumed. The highest ultimate yields of the type II elimination products and the cyclobutanol can be calculated by scaling the sum of their quantum yields to 100%, giving 88% and 12%, respectively.

One might wonder why the sum of the quantum yields is 1.04, greater than 1. This can be attributed to experimental error in the quantum yields.

49. The products appear to be the result of photochemically induced, pericyclic reactions: two retro[2+2] cycloadditions and one electrocyclic ring opening. Photochemical [2+2] cycloaddition reactions are allowed, so the reverse should also be allowed. The observed stereospecificity in the retrocycloadditions is suggestive of concerted pathways, though biradical mechanisms would also be possible, as long as biradical cleavage is faster than the conformational change required to form the other isomer.

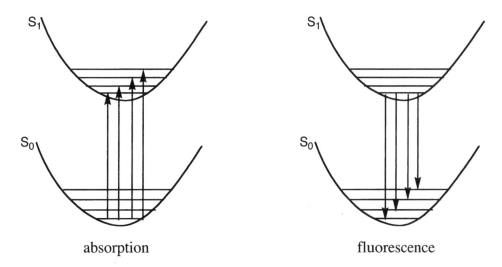

The lack of an electrocyclic product in the second (trans) case can be explained by noting that the photochemical cyclobutene opening apparently proceeds in a disrotatory fashion. A conrotatory opening from the cis isomer would produce a product that contained one *trans-*cyclohexenyl group (very strained), while a disrotatory opening can lead to the observed product. On the other hand, a conrotatory opening would be needed from the trans cyclobutene in order to produce the stable, *cis,cis* form of the butadiene product. The fact that this product is not observed is consistent with the notion that the conrotatory process is photochemically forbidden.

50. The spacings between maxima or shoulders in an absorbance spectrum reflect the vibrational levels of the excited state, since absorption generally proceeds from the ground vibrational level of S_0 to various vibrational levels of S_1. Likewise, the spacings between maxima or shoulders in an fluorescence spectrum reflect the vibrational levels of the ground state. Since bonds are often weakened upon excitation, as an electron is removed from a bonding MO and placed into an antibonding MO, vibrational energies are often decreased in an excited state. This leads to smaller spacings in the absorption spectrum relative to the fluorescence spectrum.

51. The isopropylidene groups in this cyclobutanone can serve as "loose bolts" that efficiently dissipate the excitation energy, preventing chemical changes. The parent cyclobutanone has no such way to dissipate energy.

Other factors that might contribute to the photostability of this ketone are the facts that Norrish I cleavage would produce a destabilized, vinylic radical center, that the γ-H atoms are unable to approach the carbonyl O very closely, and that the excited state, whether n,π* or π,π*, should be relatively low in energy due to the conjugation.

52. The best pathway for this reaction involves two photochemical steps: a Norrish I cleavage and a Paterno-Buchi reaction, with a hydrogen abstraction step in-between:

1. The band structure of polyacetylene based on butadiene includes four bands, one from each of the π orbitals of butadiene:

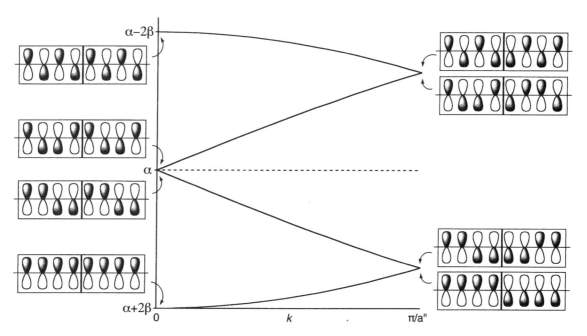

This band structure can be obtained by just "folding" the bands based on ethylene (Figure 17.4), as depicted in Figure 17.5. The band energies shown above were calculated using Eq. 17.4 and then folding twice. (A suitable answer to this exercise requires no calculation – a reasonable qualitative picture can be obtained through either consideration of the AO overlaps or through folding.)

Peierls distortion, still taken as a lengthening and shortening of alternate bonds, gives rise to the folded form of Figure 17.8, which now has a band gap:

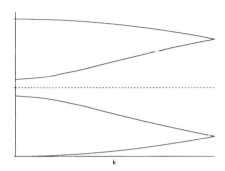

Remember that by considering a Peierls distortion we are necessarily going beyond HMO theory, since HMO treats all adjacent interactions as equivalent, but we are now considering alternating short (strong) and long (weak) interactions.

2. As suggested in step 4 at the end of Section 17.1.1, we can estimate the band energies at the zone edge by considering the energies of the corresponding MOs and the magnitude of the intercell interactions. The benzene HMO energies and coefficients are given in Section 14.3.2. For bands 1 and 3, the HMO energies are $\alpha + 2\beta$ and $\alpha + \beta$. From Figure 17.7, we can see that the zone edge intercell interactions have at least the correct signs for a degeneracy: the antibonding interactions in band 1 raise the energy from $\alpha + 2\beta$, while the bonding interactions in band 3 lower the energy from $\alpha + \beta$. Further, since the AOs involved in the intercell interaction have larger coefficients in band 3 than in band 1, the dispersion should be higher for band 3, also consistent with the band diagram.

In order to be more quantitative in evaluating the intercell interactions, let's consider Eq. 14.11:

$$E = \langle \psi | H | \psi \rangle / \langle \psi | \psi \rangle$$

For crystal orbitals at the zone center or edge, we can consider ψ to be the MO of the unit cell with the addition of one intercell interaction (say between ϕ_A and ϕ_Z) to H. All interactions in the polymer are then equally represented. Since ψ is normalized, the denominator above is equal to 1. Expansion of the numerator will include all of the terms that go into the energy of the monomer MO, plus the intercell interaction: $2c_A c_Z H_{AZ}$. (The cross term, $c_A c_Z H_{AZ}$, appears twice, since $c_A \phi_A$ and $c_Z \phi_Z$ each appear in the bra and ket. This is analogous to the two-orbital mixing of Eq. 14.40.) At the HMO level, $H_{AZ} = \beta$. Thus, we can estimate the energy of a zone center or zone edge crystal orbital by simply adding $2c_A c_Z \beta$ to the monomer MO energy.

For band 1, the HMO energy is $\alpha + 2\beta$, and the intercell interaction at the zone edge is $2\beta \left(1/\sqrt{6} \right)\left(-1/\sqrt{6} \right) = (-1/3)\beta$. So the zone-edge energy for this band is $\alpha + 2\beta - \frac{1}{3}\beta = \alpha + \frac{5}{3}\beta$.

For band 3, the HMO energy is $\alpha + \beta$, and the intercell interaction at the zone edge is $2\beta \left(2/\sqrt{12} \right)^2 = (2/3)\beta$. So the zone-edge energy for this band is $\alpha + \beta + \frac{2}{3}\beta = \alpha + \frac{5}{3}\beta$. Thus, this HMO-level method does predict an exact degeneracy at the zone edge.

3. Referring to Figure 14.32, we have two alternative ways to represent the NBMOs of cyclobutadiene: with coefficients on all C's or only on diagonal C's. The latter method is more appropriate for the current problem, since the symmetry of each NBMO is consistent with the 1,3 substitution pattern. One of these NBMOs will then have no coefficients at the ring junctions and will therefore lead, at the HMO level, to a completely flat band. The other NBMO has quite strong inter-ring interactions, leading to a band with high dispersion.

Energy calculations, as described in exercise 2, can be done to estimate the energies of each band at the zone center and zone edge, as shown in the following table. The calculations are most easily set up after drawing the zone center and zone edge crystal orbitals, as shown in the band diagram below.

Zone center	Zone edge
$E_4 = \alpha - 2\beta + 2\beta(0.5)(0.5) = \alpha - 1.5\beta$	$E_4 = \alpha - 2\beta + 2\beta(0.5)(-0.5) = \alpha - 2.5\beta$
$E_3 = \alpha + 2\beta(-0.707)(0.707) = \alpha - \beta$	$E_3 = \alpha + 2\beta(-0.707)(-0.707) = \alpha + \beta$
$E_2 = \alpha + 2\beta(0.0)(0.0) = \alpha$	$E_2 = \alpha + 2\beta(0.0)(0.0) = \alpha$
$E_1 = \alpha + 2\beta + 2\beta(0.5)(0.5) = \alpha + 2.5\beta$	$E_1 = \alpha + 2\beta + 2\beta(0.5)(-0.5) = \alpha + 1.5\beta$

We can next generate a band diagram by plotting the zone center ($k = 0$) and zone edge ($k = \pi/a$) energies and connecting the bands with curves that approximate cosine functions, as predicted by Eq. 17.4 for polyacetylene. A similar curvature is likely for any band that arises from intercell overlap of p orbitals, as long as it is not a folded band, like the doubly folded band of exercise 1.

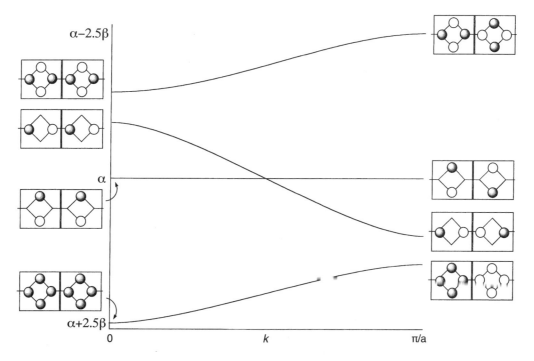

The Fermi level is α, where the two NBMO-derived bands intersect. Each of these middle bands will be half-filled, and poly(1,3-cyclobutadiene) should therefore be a conductor (a metal). This, of course, presumes that it is stable enough to exist! Like cyclobutadiene itself, the monomer units in the polymer would likely be very reactive and would tend to dimerize with units in neighboring chains.

The bands in this diagram – and those in the remaining band structure exercises below – are semiquantitative, in that the energies have been estimated at the endpoints. Using this simple method, the most important issues, such as band gap, are reasonably approximated even if the band shapes are not precise.

4. A first point is that consideration of the Jahn-Teller distortion on the band structure requires a deviation from the HMO level (as noted for the Peierls distortion in exercise 1). If we allow a shorter bond to experience a stronger interaction, we get the following band structure from the diagram in exercise 3. (The geometric distortion is exaggerated for clarity in the structures below.)

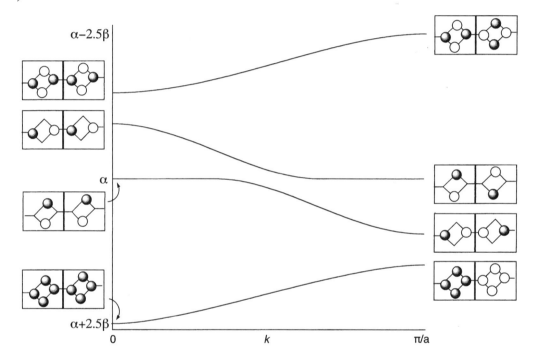

The distortion is entirely within the ring, and the effect on the highest and lowest bands is minimal due to cancellation of bond shortening and lengthening effects. At first, it would appear that the middle two bands would also be unaffected, since these do not have any intra-ring interactions. However, the distortion does have an important effect: it removes the mirror plane symmetry elements that were present in the square unit. In the diagram of exercise 3, the two NBMO bands are shown to cross. This occurs because the symmetries of the NBMOs are different with respect to the mirror planes. With the distortion, the only remaining symmetry (aside from the plane of the ring, with respect to which all of the orbitals are antisymmetric) is a C2 axis, and both NBMOs are antisymmetric with respect to this axis. Therefore, instead of crossing, the bands mix and form an avoided crossing. Note that the atoms with non-zero coefficients change from one end of these bands to the other. In the region of mixing, all four C's have non-zero coefficients, similar to the alternative depiction of the cyclobutadiene NBMOs shown in Figure 14.32.

An important point is that the structure seemingly still has no band gap, since the two NBMO bands either start or end at $E = \alpha$. Though the two bands mix most strongly in the center, they do mix to a small extent at all k values, and this leads to a small band gap. Calculation puts this band gap at $0.008\,\beta$.

5. The two structures shown for PPP are not representations of two different, degenerate structures but are resonance structures that average to a single hybrid structure. In contrast, the two structures for polyacetylene are *not* resonance structures, since this polymer really has alternating long and short bonds (due to Peierl's distortion). Thus, while polyacetylene has two degenerate structures, PPP does not. So PPP cannot support solitons.

6. High band dispersion is a result of strong interactions between monomeric units, and strong interactions should also lead to high carrier mobility.

7. We can construct the bands for poly(m-phenylene) in a similar manner to the *para* isomer (Figure 17.7), using the method outlined in exercise 3. Two comments should be made about symmetry. First, the degenerate benzene MOs should be oriented such that they are symmetric with respect to the *meta* substitution (*i.e.*, with a vertical mirror plane for the polymer with a horizontal backbone, as drawn in the diagram below). Second, since the *meta* C–C bonds are not parallel, our drawing of the polymer makes use of an up-down alternation of monomer orientation.

After sketching the endpoint crystal orbitals for the band diagram (shown below), we next estimate their energies:

<table>
<tr><td align="center">Zone center</td><td align="center">Zone edge</td></tr>
<tr><td>$E_6 = \alpha - 2\beta + 2\beta\,(1/\sqrt{6})(1/\sqrt{6}) = \alpha - 1.67\beta$</td><td>$E_6 = \alpha - 2\beta + 2\beta\,(1/\sqrt{6})(-1/\sqrt{6}) = \alpha - 2.33\beta$</td></tr>
<tr><td>$E_5 = \alpha - \beta + 2\beta(-0.5)(0.5) = \alpha - 1.50\beta$</td><td>$E_5 = \alpha - \beta + 2\beta(-0.5)(-0.5) = \alpha - 0.50\beta$</td></tr>
<tr><td>$E_4 = \alpha - \beta + 2\beta\,(1/\sqrt{12})(1/\sqrt{12}) = \alpha - 0.83\beta$</td><td>$E_4 = \alpha - \beta + 2\beta\,(1/\sqrt{12})(-1/\sqrt{12}) = \alpha - 1.17\beta$</td></tr>
<tr><td>$E_3 = \alpha + \beta + 2\beta(-0.5)(0.5) = \alpha + 0.50\beta$</td><td>$E_3 = \alpha + \beta + 2\beta(-0.5)(-0.5) = \alpha + 1.50\beta$</td></tr>
<tr><td>$E_2 = \alpha + \beta + 2\beta\,(1/\sqrt{12})(1/\sqrt{12}) = \alpha + 1.17\beta$</td><td>$E_2 = \alpha + \beta + 2\beta\,(1/\sqrt{12})(-1/\sqrt{12}) = \alpha + 0.83\beta$</td></tr>
<tr><td>$E_1 = \alpha + 2\beta + 2\beta\,(1/\sqrt{6})(1/\sqrt{6}) = \alpha + 2.33\beta$</td><td>$E_1 = \alpha + 2\beta + 2\beta\,(1/\sqrt{6})(-1/\sqrt{6}) = \alpha + 1.67\beta$</td></tr>
</table>

The band diagram can then be completed:

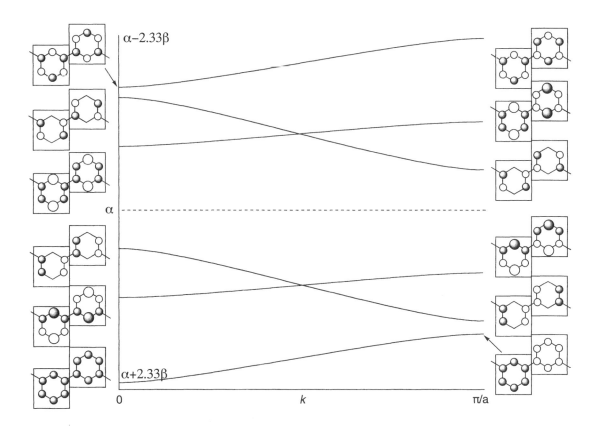

We can explain the larger band gap for the *meta* isomer by noting that the bands nearest the gap have relatively smaller dispersions, since the largest MO coefficients $(2/\sqrt{12})$ are away from the ring junctions. In contrast, the *para* isomer has one filled band and one empty band that result from overlap of the largest coefficients, giving a larger dispersion and a smaller band gap.

8. We again follow the method of drawing the zone center and zone edge crystal orbitals (shown on the diagram below), estimating the energies for these orbitals as outlined in exercise 2, and then plotting the bands by connecting the endpoints. Butadiene HMO energies and coefficients are given in Figure 14.15B. It is important to note that this polymer has two intercell interactions between each monomer pair. Our calculations are simplified if we notice that the intercell interactions have the same magnitude for each of the crystal orbitals, either positive or negative: $2\beta[(0.37)(0.60) + (0.37)(0.60)] = 0.89\beta$.

Zone center	Zone edge
$E_4 = \alpha - 1.62\beta - 0.89\beta = \alpha - 2.51\beta$	$E_4 = \alpha - 1.62\beta + 0.89\beta = \alpha - 0.73\beta$
$E_3 = \alpha - 0.62\beta - 0.89\beta = \alpha - 1.51\beta$	$E_3 = \alpha - 0.62\beta + 0.89\beta = \alpha + 0.27\beta$
$E_2 = \alpha + 0.62\beta + 0.89\beta = \alpha + 1.51\beta$	$E_2 = \alpha + 0.62\beta - 0.89\beta = \alpha - 0.27\beta$
$E_1 = \alpha + 1.62\beta + 0.89\beta = \alpha + 2.51\beta$	$E_1 = \alpha + 1.62\beta - 0.89\beta = \alpha + 0.73\beta$

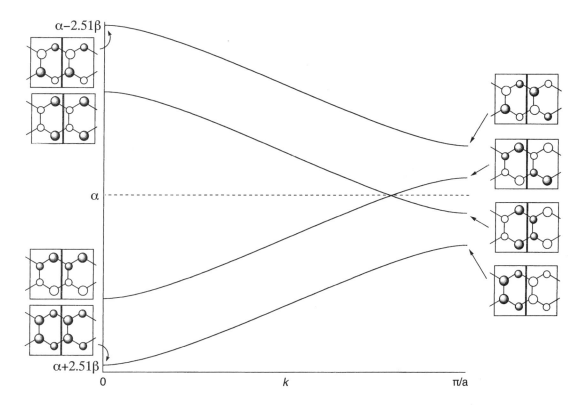

The band diagram allows the prediction of a zero band gap, since the middle two bands cross (not an avoided crossing, since the symmetries are different with respect to the horizontal mirror plane). Band 2 would be mostly filled, and band 3 would be partially filled, with the Fermi energy at α. Thus, this polymer is predicted to be a conductor.

Polyacene, like polyacetylene, has two degenerate forms that arise from a Peierls distortion. Therefore, polyacene can support solitons.

9. The limiting band energies are first estimated, and the bands are then plotted:

Zone center

$E_5 = \alpha - 1.62\beta + 2\beta(0.372)(-0.372) = \alpha - 1.90\beta$

$E_4 = \alpha - 0.97\beta + 2\beta(-0.581)(-0.581) = \alpha - 0.29\beta$

$E_3 = \alpha + 0.62\beta + 2\beta(0.602)(-0.602) = \alpha - 0.10\beta$

$E_2 = \alpha + 1.05\beta + 2\beta(-0.028)(-0.028) = \alpha + 1.05\beta$

$E_1 = \alpha + 2.02\beta + 2\beta(0.402)(0.402) = \alpha + 2.34\beta$

Zone edge

$E_5 = \alpha - 1.62\beta + 2\beta(0.372)(-0.372) = \alpha - 1.34\beta$

$E_4 = \alpha - 0.97\beta + 2\beta(-0.581)(-0.581) = \alpha - 1.65\beta$

$E_3 = \alpha + 0.62\beta + 2\beta(0.602)(-0.602) = \alpha + 1.34\beta$

$E_2 = \alpha + 1.05\beta + 2\beta(-0.028)(-0.028) = \alpha + 1.05\beta$

$E_1 = \alpha + 2.02\beta + 2\beta(0.402)(0.402) = \alpha + 1.70\beta$

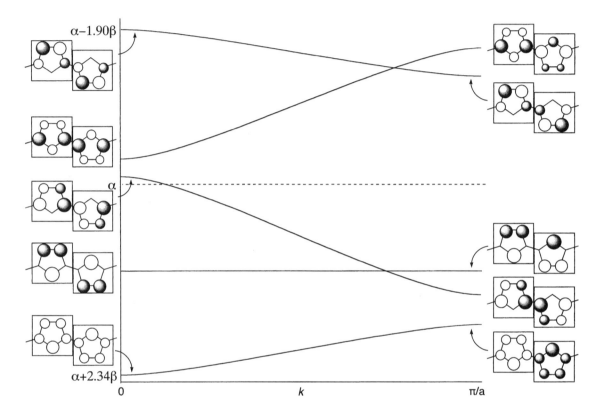

The band gap in polythiophene is quite small (0.19β), because the two bands with the largest dispersions happen to derive from the HOMO and LUMO of thiophene.

10. The band diagram for polyisothianaphthalene should be quite similar to that for polythiophene, since the benzene rings are fused away from the polymer connection points. The important band orbitals to consider with respect to the band gap are the zone center orbitals for bands 3 and 4. These orbitals have the correct symmetries to interact with the butadiene HOMO and LUMO, respectively, as shown in the following MO diagram. These interactions serve to push these band orbitals closer together, reducing the band gap. Note that the band orbitals are represented by only one monomer unit, which could represent any orbital in the given band, but the zone center energies are used in the diagram.

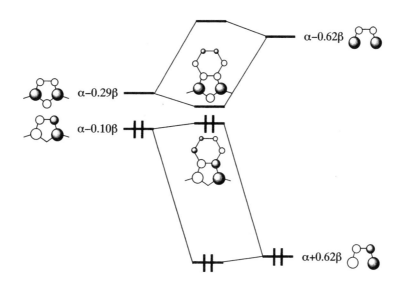

$\alpha-0.62\beta$

$\alpha-0.29\beta$

$\alpha-0.10\beta$

$\alpha+0.62\beta$

11. The zone center and zone edge energies are calculated as in the previous solutions by using the HMO energies and orbitals of Figure 14.11, and the bands are then plotted:

Zone center	Zone edge
$E_6 = \alpha - 1.86\beta + 2\beta(-0.44)(-0.44) = \alpha - 1.47\beta$	$E_6 = \alpha - 1.86\beta + 2\beta(-0.44)(0.44) = \alpha - 2.25\beta$
$E_5 = \alpha - 1.62\beta + 2\beta(-0.37)(0.37) = \alpha - 1.89\beta$	$E_5 = \alpha - 1.62\beta + 2\beta(-0.37)(-0.37) = \alpha - 1.35\beta$
$E_4 = \alpha - 0.25\beta + 2\beta(-0.35)(-0.35) = \alpha - 0.01\beta$	$E_4 = \alpha - 0.25\beta + 2\beta(-0.35)(0.35) = \alpha - 0.50\beta$
$E_3 = \alpha + 0.62\beta + 2\beta(0.60)(-0.60) = \alpha - 0.10\beta$	$E_3 = \alpha + 0.62\beta + 2\beta(0.60)(0.60) = \alpha + 1.34\beta$
$E_2 = \alpha + 1.00\beta + 2\beta(0.00)(0.00) = \alpha + 1.00\beta$	$E_2 = \alpha + 1.00\beta + 2\beta(0.00)(0.00) = \alpha + 1.00\beta$
$E_1 = \alpha + 2.12\beta + 2\beta(0.43)(0.43) = \alpha + 2.49\beta$	$E_1 = \alpha + 2.12\beta + 2\beta(0.43)(-0.43) = \alpha + 1.75\beta$

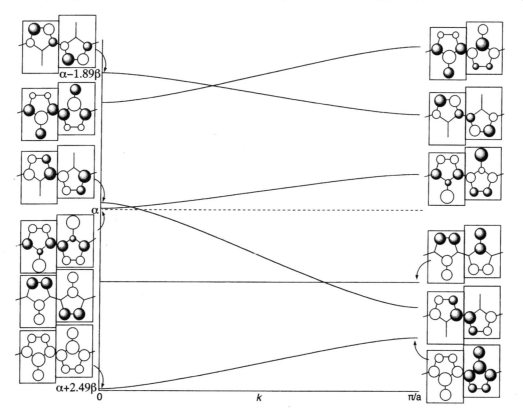

Since the unit cell has six π electrons, three of the bands are filled. A slight crossing exists between the third and fourth bands, resulting in a zero band gap. The energies of these bands differ by only 0.09β at the zone center – a near degeneracy. This is a situation similar to that for polyacetylene (even in the butadiene-like frontier bands – compare to exercise 1), so a similar Peierls distortion could lead to the observed band gap of 0.9 eV, a value not far from that for polyacetylene (1.5 eV).

12. Both types of structural variability concern the isomeric relationships that are possible as monomeric units are linked together. These include regioisomeric linkages arising from ring orientation: head-to-tail vs. head-to-head and tail-to-tail. Another type of variability comes from diastereotopic relationships of the side chains: *R,S* vs. *R,R* and *S,S*. So not counting enantiomeric linkages, there are six different types of linkages in this polymer.

13. The following reaction with base gives directly a highly conjugated intermediate:

This non-aromatic, quinodal structure would be expected to be reactive and to have a red-shifted absorbance relative to a benzene derivative. Polymerization can involve either radical or anion chain mechanisms. The radical mechanism:

Initiation (two possibilities):

Propagation:

Anionic initiation:

The propagation step is analogous to the radical propagation step with an anionic center in place of the radical center. Whether polymerization goes by a radical or anionic mechanism, base-induced 1,2-elimination would then give in-chain alkenes, leading to the conjugated polymer.

A simple experiment to distinguish these polymerization mechanisms would be to measure the rate of intermediate decay (using spectroscopy) in the presence and absence of a radical inhibitor. Alternatively, the rate of polymer formation could be measured gravimetrically by taking aliquots. If the inhibitor slowed the polymerization, at least part of the reaction could be attributed to the radical pathway. If the inhibitor had no effect on the rate, the radical pathway would be eliminated.

14. An antiferromagnet can be distinguished from a diamagnet by measuring the temperature dependence of magnetization. At high enough temperature, an antiferromagnet, with fairly weak coupling between the unpaired spins, will become a paramagnet.

15. In a infinite, one-dimensional ferromagnet, each and every pair of adjacent spins must be aligned parallel and not antiparallel. We might think that this would be possible if the energetic preference for high-spin coupling were high enough. However, since the number of such pairings is infinite, at any temperature above 0K the entropic term in ΔG ensures the presence of some low-spin pairings. A more practical consideration leads to the same conclusion. Even in a finite chain, perfect order cannot be expected above 0K, and we should expect that any material will have defects, such that in some locations the spin alignment will break down. These defects could be statistical (entropic) in nature, as noted above or could be associated with physical differences such as conformational variations or a missing spin. In this way, parts of the sample become isolated from each other, and the preference for up or down spin may be reversed. So even if the spin coupling is strong, the presence of a very few defects will cause the net magnetization to average to zero. In a two-dimensional ferromagnet, defects will not necessarily cause reversal of the spin direction since the preference can be enforced by more than two neighbors.

16. By incorporating rings, the number of spin-coupling paths increase. A defect that disrupts one coupling path will then have less impact. For example:

Consider the right example with a defect: a CH_2 in place of a carbene (at right below). Even a central defect does not isolate any of the other carbenes. Comparing this with the related structure from Figure 17.19 (at left below), we can see that a similar defect separates the remaining carbene sites into two distinct sets.

17. The mechanism of ferromagnetic coupling in polytrityl systems depends on electronic interactions through the π system. Even though perchlorination adds stability (or more precisely, persistence), it also reduces π interactions through steric inhibition of resonance. (The perchlorotrityl units will be much less planar.)

18. The crystal structure of quartz is non-centrosymmetric, allowing for the possibility of SHG and other second-order NLO effects.

19. A polymer used for poling must first have T_g somewhat above ambient temperature (assuming the material will be used at room temperature). If T_g is too close to ambient temperature, the chromophore alignment will not be very securely locked, and the alignment will degrade too quickly. If the T_g is too high, the poling process will require high temperatures that might degrade the chromophore. The best polymer will be one that has a highly temperature dependent viscosity.

   Another important requirement is that the polymer be transparent in the wavelength range of interest. This relates not only to light absorption, but also scattering. Thus a highly amorphous polymer is preferable to one with significant crystallinity, which can lead to scattering centers.

20. We can show that the bipolaron has more bonding interactions than two polarons just by counting π bonds in the structures of Figure 17.10B. The structure with two polarons has 25 π bonds, and the structure with a bipolaron has 26. One can easily see why this is the case: the two radical centers of the polarons have combined to form an additional bond in the bipolaron.

21. Starting from 1,3,5-tribromobenzene, we can produce the tris(Grignard) reagent by reaction with Mg metal. The G1 dendrimer could then be formed by adding 3,5,3',5'-tetrabromobenzophenone followed by protonation.

Repeating the first two steps, withholding protonation until the end, would give the G2 dendrimer, and higher generations could likewise be produced. The G2 dendrimer is already impressive!

22. In both the linear and dendrimeric systems, the effect of a single defect depends on the location. In either system, the defect will have the minimum effect if it happens to be an "outside" site. The probabilities of this are quite different: 9.5% for the linear system and 57% for the dendrimeric system. In the dendrimeric system, the largest possible effect is that only two thirds of the sites will remain connected, and the probability of this is 14%. In the linear system, it is possible that less than half the sites will remain connected, and the probability that two thirds or fewer of the sites remain connected is 33%. The linear system is clearly more sensitive, on average, to the presence of a defect.

23. The presence of the electronegative O in *m*-quinodimethane suggests that the following zwitterionic resonance structures might be important in this case:

One might think that the singlet state would be stabilized by the participation of these structures, which by themselves would very clearly favor the singlet state.

24. All of the biradicals shown are alternant hydrocarbons (AHs), since they have only even-membered rings. Therefore, we can use the */non-* method: any AH with equal numbers of starred and non-starred atoms will have disjoint NBMOs and a preference for the singlet state, and any AH with non-equal numbers will have non-disjoint NBMOs and a preference for the triplet state.

| # starred-# non-starred: | 7-7 = 0 | 8-6 = 2 | 7-7 = 0 |
| spin preference: | singlet | triplet | singlet |

| # starred-# non-starred: | 4-2 = 2 | 5-5 = 0 |
| spin preference: | triplet | singlet |

| # starred-# non-starred: | 7-5 = 2 | 6-6 = 0 |
| spin preference: | triplet | singlet |

| # starred-# non-starred: | 7-5 = 2 | 7-5 = 2 |
| spin preference: | triplet | triplet |

25. By attaching a non-starred atom of a new branch to a starred atom of the main chain, we can preserve the growing excess of starred atoms and ensure high-spin coupling:

26. Since the triradical is an alternant hydrocarbon, we can use the */non-* method to determine whether it is high spin:

There are three more starred atoms than non-starred (24 relative to 21), so the triradical will indeed be high spin.

27. Since solids are stiffer at lower temperatures, and since superconductivity is a low-temperature phenomenon, one might infer that lattice stiffness promotes superconductivity. Indeed, the reason that superconductivity breaks down above T_c is that lattice vibrations other than those induced by the lattice pairs interfere. Therefore, a solid of high enough stiffness to prevent undesirable lattice modes should be able avoid this problem until higher temperatures, thus raising T_c. One should realize, however, that understanding the structure/property effects in superconducting materials is a very challenging task, and a variety of behaviors with respect to lattice stiffness have been observed.

28. One possible pair of Suzuki reactants is

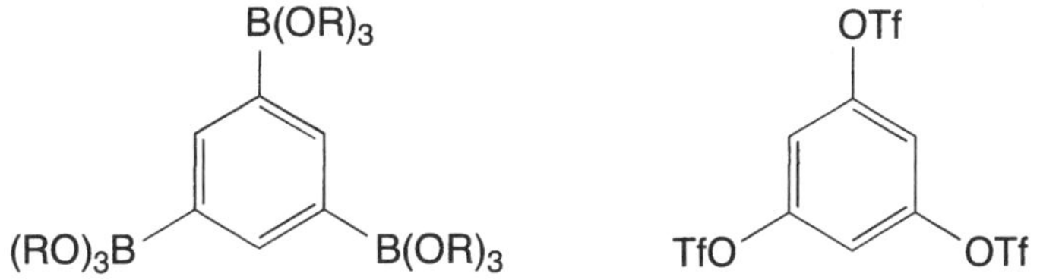

Using the tris(boronic ester) as the core and both molecules in alternating fashion to build the arms, the dendrimers after the first and second steps would look like:

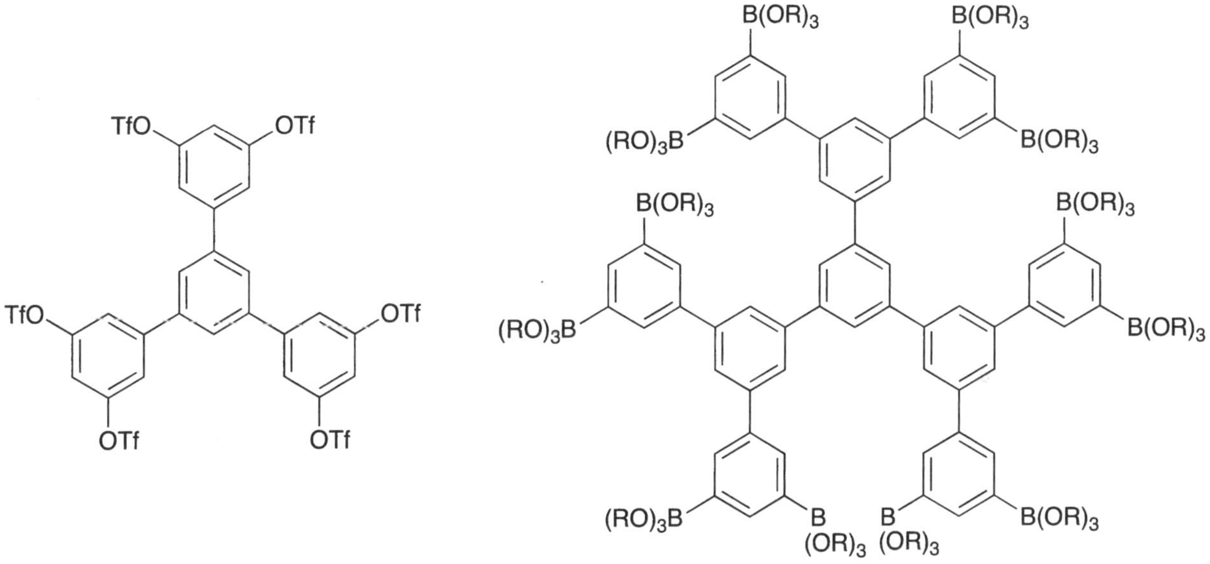

However, this approach has potential selectivity problems. In each step, we want a trifunctional molecule to react only at one of its reactive sites. An undesired reaction would be the bridging of two cores. Though we can increase the yield of the desired product by using significant excesses of the trifunctional reagent, this approach is wasteful and requires a potentially difficult separation after each step. A related but more strategic approach would use, for example, the following two components, the first as core and the second as arm building block:

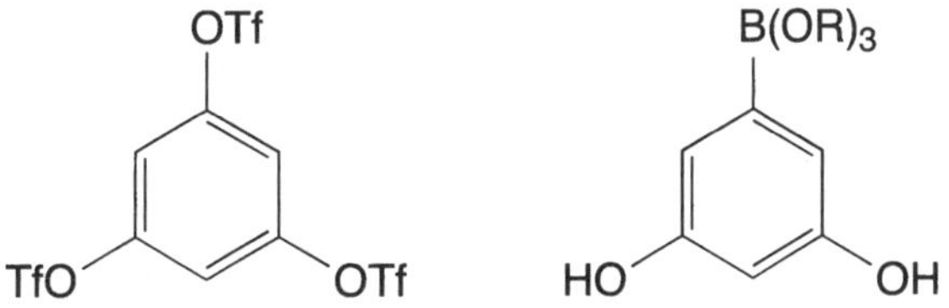

Since the arm building block has only one reactive group, the selectivity problems are eliminated. After each arm-linking reaction, the OH groups can then be converted to triflate groups, and the molecule will be ready for the next arm-linking reaction.

A P P E N D I X

5

Pushing Electrons

S O L U T I O N S   T O   E X E R C I S E S

1.

	Electron sources	Electron sinks
A.	O lone pairs	C's attached to O (not very reactive)
B.	(Cl lone pairs are very unreactive)	C attached to Cl
C.	O lone pairs	somewhat acidic H (on O) C attached to O (not very reactive)
D.	O lone pairs π electrons (less reactive)	C of carbonyl
E.	π electrons	(alkene C's are very unreactive)
F.	N lone pair π electrons (less reactive)	C of imine
G.	(O lone pair, π electrons are very unreactive)	acidic H C of protonated carbonyl

2. A.

B.

This mechanism includes a second equivalent of amine that acts as a base. Mechanisms are often shown with less detail, for example without showing the identity of the base. Thus, you will often see mechanisms with "-H⁺" over the arrow, even though we know that there is always a base that removes the proton (amine, solvent, Cl⁻, etc.). This is sometimes a convenient convention, since the identity of the base is often unclear. For example, in the present case, unless the amine is present in high concentration, the base that removes the proton from the protonated amide is more likely to be the solvent. (The more basic amine will still end up with the proton eventually.) Issues such as these are often considered minor mechanistic details that need not be written, but they sometimes have important consequences and should be considered even if not written.

C.

D.

This mechanism and others like it are often shortened by combining the second and third steps. Instead of using an external base (or acid) component, the proton can be transferred intramolecularly, with "~H⁺" or "H⁺ shift" written over the arrow. While this is a convenient shortcut, experiments have suggested that such intramolecular transfers are relatively slow in cases such as this that involve a four-membered ring transition state. See Section 10.15.1.

E.

3. A.

B.

C.

D.

E.

F.

G.

H.

I.

J.

K.

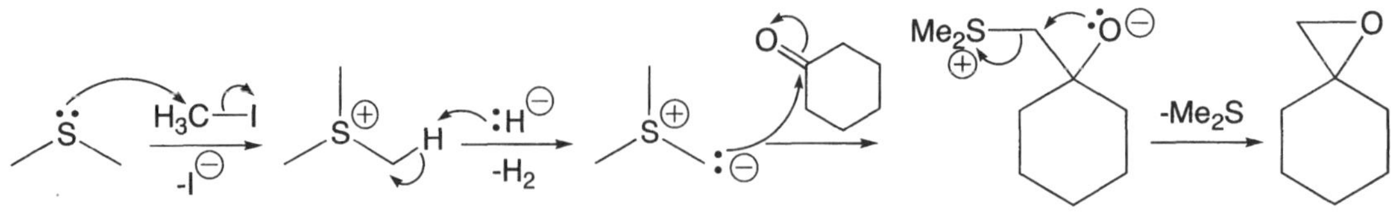

The zwitterion formed in the second step is a sulfur ylide.

L.

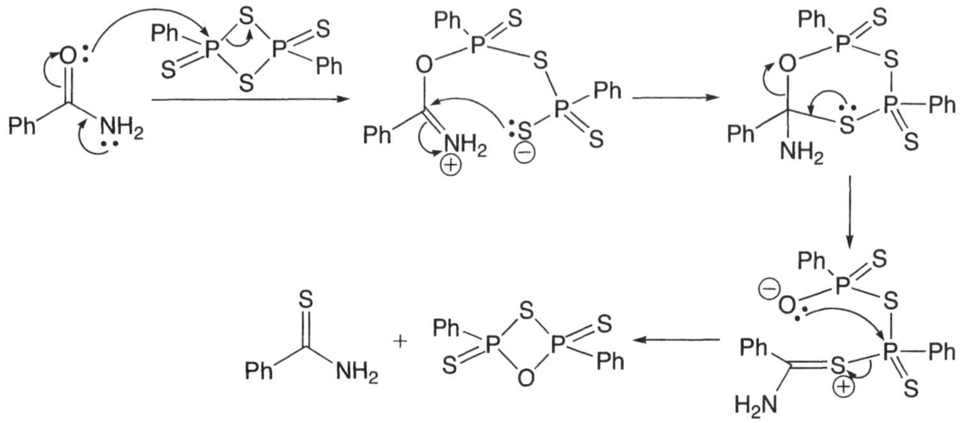

M.

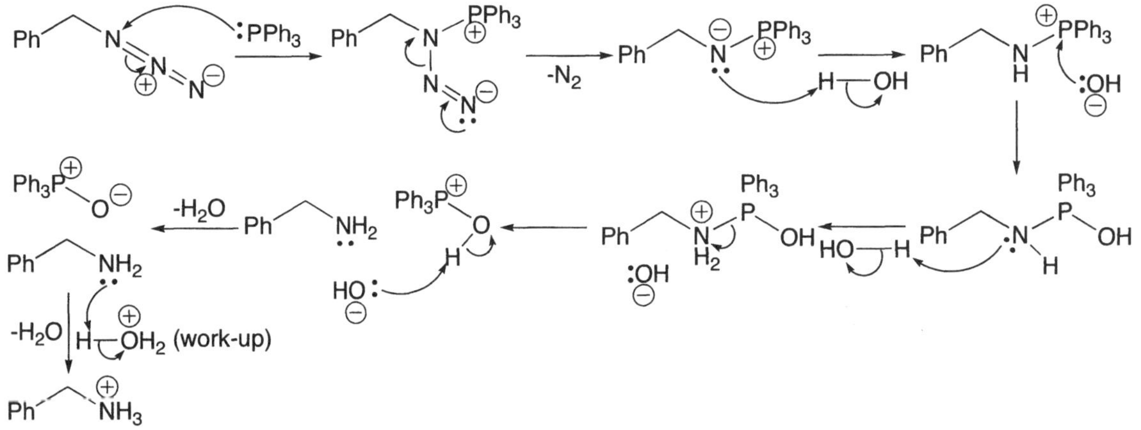

N.

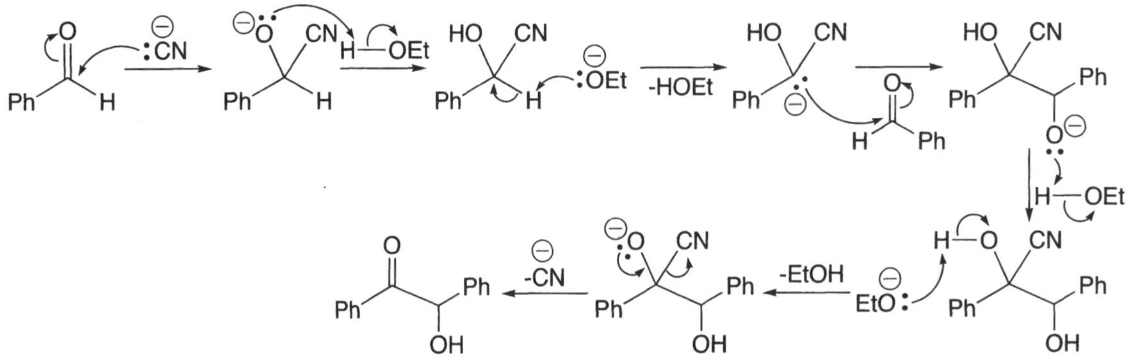

The carbanion formed in the third step is stabilized both by the cyano and phenyl groups.

O.

Note that the elimination in the second-to-last step could be drawn without protonation at N, but it should be facilitated by the positive charge on N.

P. The beginning of each of the three steps is noted on the arrows.

The second three-membered ring opening is drawn as concerted with loss of water. While this could be drawn in two steps, with opening following loss of water, the azacyclopropyl cation intermediate, a nitrenium ion, would be expected to be very short-lived if indeed it is an intermediate. The opening not only releases strain but also forms a resonance-stabilized cation. Two steps later, the isomerization of an imine is drawn as occurring with general base catalysis by water in order to avoid another nitrenium ion.

4. All three reactions are free-radical chain processes. The initiation process can be relatively inefficient, since the propagation steps can cycle hundreds or thousands of times for each initiation. The propagation steps account for product formation, and these steps should sum to the observed reaction. Termination steps destroy propagating radicals. Though free radical chain mechanisms can be drawn in a similar way to two-electron processes such as those in exercises 2 and 3, writing each step separately is generally clearer for these reactions.

A. Initiation:

Propagation:

Termination:

Many other termination steps are possible.

B. Initiation: production of free radicals, In•, from impurities or added initiator.

Propagation:

Termination (for example):

C. Initiation:

$$Br_2 \longrightarrow 2\ Br\cdot$$

Propagation:

Termination (for example):

Thomas
O'Connor
710
Latimer